Northrop Grumman
B-2
Spirit
An Illustrated History

Bill Holder

Schiffer Military/Aviation History
Atglen, PA

Acknowledgments

1. Edward L. Smith, Manager, Corporate Northrop Grumman Public Information
2. Colonel Steve Heaps, B-2 Deputy Program Manager
3. Sergeant Dave Johnson, B-2 Systems Project Office
4. Michael Tull, Boeing Company Office of Information

Book Design by Ian Robertson.

Copyright © 1998 by Bill Holder.
Library of Congress Catalog Number: 98-84395

All rights reserved. No part of this work may be reproduced or used in any forms or by any means – graphic, electronic or mechanical, including photocopying or information storage and retrieval systems – without written permission from the copyright holder.

Printed in China.
ISBN: 0-7643-0591-3

We are interested in hearing from authors with book ideas on related topics.

Published by Schiffer Publishing Ltd.
77 Lower Valley Road
Atglen, PA 19310
Phone: (610) 593-1777
FAX: (610) 593-2002
E-mail: schifferbk@aol.com
Please write for a free catalog.
This book may be purchased from the publisher.
Please include $3.95 postage.
Try your bookstore first.

Printed in Hong Kong.

Table of Contents

Forewords:
Lt. Col Tony Grady (Ret)
Former B-2 Squadron Commander (1996-1997) 4
Lt. Col Jeffrey Smith
B-2 System Program Office (1994-1997) 5

Chapter 1.	An Introduction	6
Chapter 2.	The First Flying Wings	14
Chapter 3.	Early B-2 Design and Development	24
Chapter 4.	B-2 Parts and Pieces	30
Chapter 5.	B-2 Production	44
Chapter 6.	B-2 Ground and Flight Test	50
Chapter 7.	Flying the B-2	60
Chapter 8.	B-2 Operational Service	66

Foreword

In a few short pages, Bill Holder has articulated the salient points of perhaps the most successful test program of any weapon system that the United States has ever fielded. Truly, the B-2 is impressive. However, that is not the real success story here. The real story is hidden in the name of the B-2, the Spirit. This name captures the essence of the real success of this program. It was the spirit of the visionary Jack Northrop who championed the innovative Flying Wing design when others were skeptical. He even kept the design alive after the Air Force canceled the original Flying Wing program. It was only fitting that he was allowed to see a scale model of the B-2 before his death. It was also the spirit of Glenn Edwards, the daring test pilot who met his demise running the YB-35 through its paces when the first Flying Wings darkened the skies over Edwards Air Force Base. It is fitting that the premier flight test facility in the free world now bears his name as a lasting testament. It was the spirit of the Northrop, and later the Northrop-Grumman partnership, that boldly planned and executed the enormous B-2 program that delivered the aircraft. It was the spirit of the contractor, operational, and flight test triumvirate who carried out the test program. Together, they discovered problems and determined when they were solved. It is the spirit of future combat pilots who daily hone their skills, perfecting the procedures and tactics necessary to fly the B-2 in combat if called upon by the national command authorities to achieve the nation's strategic objectives. However, the real heroes and heroines are the personnel who worked on the B-2 program.

There were 40,000 people over a 15 year period who spent long nights without sleep, away from their families for weeks and sometimes months at a time in total secrecy solving the myriad of technical problems that threatened to undue the aircraft at every turn. They will never have their names in lights, they will never by repaid in worldly terms. However, they possess satisfaction that few ever experience throughout their lifetime. They gave of themselves to a program that will live on, directly due to their efforts. It is to them that the country will always be indebted. However, for any of them who witnessed the public response to the B-2 when it flew over the Rose Bowl parade in Pasadena in January 1997, were partly satisfied. The parade literally came to a standstill as the Spirit suddenly appeared, flew over the crowd, and then disappeared. The appearance of the aircraft took the crowd's breath away. It was the effort of the outstanding B-2 personnel group that produced the world's most complicated and lethal weapon system that can strike anywhere on the face of the earth, anywhere, anytime.

This book is written for the layman. It is not a technical treatment, nor does it go into excruciating detail. It is the story of a successful program. It gives the reader an overview of how the B-2 came into existence. So sit back and relax. Let your very able literary pilot Bill Holder take you on a guided tour of the contours of the Spirit. When you land, I hope your spirit is enhanced.

Lt. Col Tony Grady (Retired)
Former B-2 Squadron Commander
(1996-1997)

Foreword

The B-2 is revolutionary. The Spirit is capable of delivering firepower that can not be matched by any other weapon system in the Air Force inventory. Operationally, it has changed the concept of strategic bombing. Its ability to destroy sixteen targets on a single pass prompted Former Air Force Chief of Staff Ronal Fogleman to say, "We are beginning to change our thinking from how many aircraft it takes to destroy one taget, to how many targets we can destroy with one aircraft."

This technological marvel is the fruits of the labor of a fantastic government-industry team. Much of the early work on the program was shrouded in secrecy. Information was compartmentalized so that, for example, those working on the wings knew nothing of the work on the engines. Indeed, the initial teams were small; however, the accomplishments were mighty. At the peak of the program, the industry team was comprised of over 12,000 companies in 48 states. These companies were responsible for developing over 900 new materials and manufacturing processes.

I am proud to say that I was a member of this team. Proud to say that I witnessed the fielding of an aircraft that stands ready to lead the way in a revolution in military affairs. The Desert Storm Joint Forces Air Component Commander, Gen Charles Horner, recognized this when he said, "I returned from the Gulf convinced that tomorrow's air commanders required—and would indeed have—a fleet of sixty or more long-range stealthy bombers."

Lt Col Jeff Smith
B-2 System Program Office
(1994-1997)

Chapter 1:
An Introduction

The mission requirements for the new plane were substantial to say the least. In the early 1980s, the Air Force released the following statement that gave an idea what the B-2 was expected to accomplish:

"The plane shall provide the capability to conduct missions across the spectrum of conflict, including general nuclear war, conventional conflict, and peacetime/crisis situations."

The skeptics of the plane in recent years have jumped on the nuclear aspects of the mission, stating that the Cold War environment in which the plane was developed is no longer in phase with the world situation in the 1990s. With the demise of the Soviet Union and the supposed lessening of the nuclear threat, these critics emphasize that this plane is no longer appropriate.

It is a sight that is hard to describe. The first time most people see the majestic B-2 Spirit, it's something that looks out of this world. It is probably the most publicized aircraft development in aviation history. (USAF Photo)

The B-2 is probably more majestic in flight than when it is sitting on the ground. Visible is that startling wing design that carries that weird trailing edge shape. (USAF Photo)

The penetration of Soviet air defenses was indeed one of the serious design criteria for the bomber. During the 1980s, the B-2 was played against the USSR environment and determined the vulnerability of the plane against many scenarios. It was determined in those investigations that even though the B-2 probably would not be completely invisible, it would still have effectiveness far exceeding current bombers.

It is interesting to note that the B-2 design and development brought forth a response from the Soviets themselves. Although many Soviet authorities acknowledged that the B-2 was a formidable weapon system, Defense Minister Dmitri Yazov said, "The stealth bomber is not considered to be a challenge."

But in the late 1990s, there were also those that felt that the B-2 could have been the most important weapon for the next century. It was pointed out that the 35 year life cycle costs for the B-2 were considerably less than a fighter wing or carrier. And when the bomb load capabilities of the fighters are compared with the B-2's, the comparison, according to the pro B-2 cadre, makes the Flying Wing bomber look even better.

The B-2 also has the advantage of being able to respond worldwide and deliver 16 tons of weapons to just about any type of ground target. With its range, the B-2 can operate from bases outside the range of enemy weapons of mass destruction.

The modern bat-winged B-2 is NOT the first American bomber to carry the low-number name. This early Martin B-2 light bomber was a far cry from its current namesake, but it has to be remembered as carrying the famous designation. (USAF Photo)

An Introduction 7

This group of the vintage B-2s in flight might not be as sleek and stunning as its grandson, but for its time it was one advanced machine. (USAF Photo)

The trio of bombers that constitute the USAF bombing force for the next century. The evolution is evident from the 1950s vintage B-52, followed by the B-1, to the futuristic B-2 Spirit. (USAF Photo)

The B-2: Now More Than Ever

Through its long development period, there were many that wanted promote the program. Here is one of a number of stickers that floated around through the plane's many contractors. (Northrop Sticker)

The B-2 in its early days has proved that it can meet the requirements asked of it. But there is that huge per-plane cost that dwarfs air systems of the past. Depending on the source of the unit price, the numbers approach $1 billion per copy. That brings forth the con argument of whether a commander would want to risk such an expensive bomber to attack a possibly insignificant target?

Then, there is that strange shape that has skeptics wondering what in the world this beast is. The world has become used to bombers with long slender fuselages, something that the B-2 definitely does not have.

But if you were to believe the data that was being dispensed by both the contractor and the Air Force, you would have thought that the B-2 could not be lived without. First, it was argued, future conflicts are likely to occur in areas where there are no forward deployed forces. The B-2 has the intercontinental range to operate against targets in the world on short notice with no forward basing support required.

The B-2 is also the first heavy bomber ever to be deployed with precision conventional weapons as a part of its weapon suite. Also, the B-2 offers that unique stealth technology, enabling it to operate autonomously and with near-impunity against air defenses anywhere in the world. The system is also the ideal for destroying selected critical targets while minimizing U.S. casualties.

The B-2's capabilities would also be compatible with large scale conflicts where advanced air defense systems existed. The B-2 would have the capability to attack the air defenses,

The official logo of the Northrop Grumman/United States Air Force B-2 Team. (USAF Logo)

Not only were there many deliberations concerning whether the B-2 would be produced in the first place, but then how many would be produced. What started off as a production run at 132 finally amounted to only 21 of the advanced bomber. (Northrop Sticker)

This Boeing promotional photo shows the evolution of bombers that the company has been involved with through the years. The lower shaded model is a 1930s B-9 that was a revolutionary design in that it was faster than the fighters of the time. And of course, that's the B-2 up high which could well be the last of the manned bombers. (Boeing Photo)

knock down those defenses for non-stealthy aircraft, and so disrupt the enemy advance since U.S. forces would not have time to deploy and repel the attack. Finally, because the B-2 is stealthy, it requires no other aircraft for protection and puts no other air crews at risk.

The argument of whether there are enough B-2s built, with the probable number being 21, or whether it should ever have been built in the first place will be settled deep into the next century. Quite honestly, there is just no way to accurately predict the state of the world in say, 2020, and what place the B-2 might be playing.

About all that can be projected is that it will be the plane that is holding together the peace of the world with its still-significant capabilities. Or its capabilities might have been completely eroded by technology or found to be wanting for new capabilities in a changing world.

Will it be standing on guard on a ramp with the Air Combat Command (or its follow-on organization), or will it be parked at Davis-Monthan AFB awaiting its final disposal?

Nobody can predict for sure, but the purpose of this book is to provide you with the background, design, and development period, production, testing, and initial operational capabilities of the system.

As can be seen, the political and military turmoil, great complexity and capabilities of the plane, the great amount of testing (both ground and airborne), the first public showing for the world to see, and the first time the B-2 took to the air all contribute to a great story to tell—here is that story.

When the B-2 first saw the light of day, it received world-wide coverage as the world was amazed at the look of this strange new bomber. (USAF Photo)

An Introduction 11

Guess there is no doubt about the trailing edge shape of the Flying Wing bomber which is clearly outlined by the B-2's shadow on the runway below. (USAF Photo)

Prime B-2 contractor Northrop Grumman certainly wasn't overdoing it when it described the B-2 as "So advanced it has actually changed thinking." (Northrop Grumman Photo)

Each of the operational B-2's have been personalized with a "Spirit of xxxxx" name. Because of Ohio's involvement in the program, one of the planes was given the "Spirit of Ohio" name tag. Here is the emblem for the honor on this flag which was in position for the ceremony at Wright Patterson Air Force Base in 1997. (Phil Kunz Photo)

An Introduction 13

Chapter 2:
The First Flying Wings

A lot has been made about the futuristic shape of the B-2 bomber, looking like a 21st century design. But many are aware that the amazing shape has been around, and even taken to the air, in other forms throughout history.

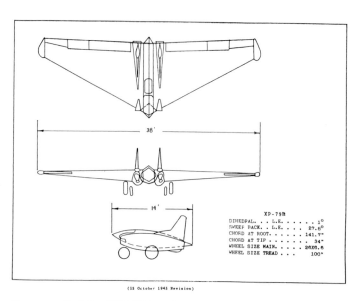

It was called the XP-79B and was one of the early prototype flying vehicles that tested the Flying Wing concept. Note that there is a short fuselage of sorts with a pair of vertical stabilizers. (USAF Photo)

The shape intrigued early engineers as far back as before World War II, the reason being the great stability of the design. But during the 1910s and 1920s, the greater complexity associated with the shape did not warrant the extra effort.

There was one early World War I design, a biplane that sported a pair of wings that looked like a pair of B-2s stacked on top of each other. Most of the Flying Wing designs, though, were not really true flying (Bi) wings, as there were many times a very small fuselage in the concepts.

During the 1930s, the interest in the Flying Wing continued, but none would reach production. Now, though, they were starting to look more like the B-2 design. Like their "normal" aircraft designs, the bi-wing concepts were gone, with the cockpits now being incorporated as a protrusion on top of the upper wing surface.

Power for these early, mostly single pilot, wings came from a pusher-engine concept, with the engine located on the centerline of the airframe. Some of the most advanced wings were developed by the Horten brothers from Germany. The experimentation started with gliders and culminated with prop-powered flight vehicles.

Some of these early concepts blended into another modern concept, the delta wing. The Lippisch Delta 1 was called a Flying Wing, but it carried a look closer to that of an F-106 Delta Dart than a B-2. It is interesting how many of our modern airframe designs had such early roots.

Of course, when discussions of early American Flying Wing research are held, the name of John K. "Jack" Northrop receives a well-deserved top billing. Northrop's fertile mind was behind the program from the beginning. He was rewarded in 1939 for his design efforts, being appointed the President and Chief Engineer of the company carrying his name.

It is generally recognized that the early Northrop wings were really the first to completely encase all the components inside the wing, making it a completely smooth surface and greatly increasing its efficiency.

Northrop's Flying Wing work began in the 1920s, and during the late part of that decade he came up with a wing

A view of the XP-59B shows the engines carried very close to the centerline of the design. Interestingly, the first actual Flying Wings did not carry the large twin vertical stabilizers. (USAF Photo)

design that did not possess the conventional fuselage. And, just like the B-2 of today, the cockpit was contained within the thick leading edge. There were also control techniques, certainly advanced for the time period, that would find utilization in later Northrop wing designs of the 1940s.

In the 1930s, the Northrop evolution would continue with the construction of the N-1M vehicle, which sported a number of revolutionary devices, including a unique tricycle landing gear, two reciprocating engines driving pusher props, and trailing edge rudders which could also be used as drag brakes. But more importantly, it was a pure wing design which featured drooping wing tips. It was manned by a single pilot.

The N-1M first flew in 1940 and completed its flight test program in 1945. It was then transferred to the Air Force Museum at Wright Patterson AFB, and then on to the Smithsonian Institution.

Two other Flying Wing testbed vehicles, the MX-324 and MX-334, would also be built. Designed for a proposed advanced rocket propelled fighter, the MX-334 was actually powered by a small rocket engine. The XP-79B fighter, which would evolve from the research, crashed on its first flight, but investigation revealed the fault was not from its Flying Wing design. The pilot was killed in the accident.

Next came a family of four N-9M test vehicles, again with a single pilot, which were built to study stability and control

Although the NM-9 flying vehicle was categorized as a Flying Wing concept, the vehicle didn't look that much like a true Flying Wing with long rear booms and a tail. The powerplant was a pusher prop located on the centerline of the plane. (USAF Photo)

The massiveness of the XB-35 is clearly evident when viewing the great size of its components seen here in construction at the Northrop Company. It's final shape would be extremely close to the B-2 which would evolve many decades later. (USAF Photo)

The First Flying Wings 15

The XB-35 didn't make it into production until after World War II, shown here in 1946. Construction of the leading edge of the plane is evident here. (USAF Photo)

It is hard to believe that such a modern Flying Wing bomber was produced over a half-century ago, but that was the case at Northrop Corporation in 1946. Unfortunately, the design would never be series produced. (USAF Photo)

characteristics. Another pilot would pay the ultimate price for the research, which would be used to directly support work on the next Flying Wing aircraft, a "real" aircraft, with the B-35 bomber. In fact, the N-9M could probably have been considered a subscale prototype of the soon-to-follow B-35 project.

The N-9M vehicles were constructed of a welded steel tube center with the exterior covered by wood. With a 60-foot wingspan, the model resembled a true Flying Wing in every sense of the definition. The N-9M weighed in at about 7,100 pounds, with the main difference between it and the B-35 bomber to follow being the number of engines. The N-9M had only a pair, while the B-35 would sport four. Even so, the

A rear view of the XB-35 prototype shows the details of the conventional trailing edge flaps it carried. The eight props and the rear of the crew module extended rearward past the rear constraints of the aircraft. (USAF Photo)

The massive tri-cycle landing gear of the XB-35 is clearly visible as the prototype takes to the air. The wheels of the gear appear to be oversized as compared to the size of the plane. (USAF Photo)

You might be surprised to learn that this large area with heavily padded benches is the crew area of the B-35 bomber. (USAF Photo)

The center crew module reached out slightly below the confines of the lower wing surface. (USAF Photo)

research results provided by the N-9M vehicles contributed greatly to the success of the overall B-35 program.

The B-35 program, which would initially carry the XB-35 designation, was born from a May 1941 requirement from the Army Air Force to Northrop that requested the company to study the possibility of an advanced bomber with an 8,000 mile range, 250 mile per hour cruising speed, and a 40,000 foot ceiling. They were all pretty demanding requirements for the technology of the time period.

The contract actually also included the financing of the N-9M. The contract also called for one XB-35 mockup, engineering data on the unique design, and an option for one additional plane.

The massive-for-the-time $2.9 million contract was, at Northrop's request, a cost-plus-fixed-fee type, because the project was going to require funds in excess of those available to the company for experimental purposes. Also, it was estimated by Northrop that the cost of materiel and labor would significantly rise before November 1943, when the XB-35 was scheduled for delivery.

It is interesting to note that had the program met that requirement, the plane might have been available to participate in World War II. An interesting thought!

Note the engine exhaust streaks that traverse the underside of the B-35 bomber. (USAF Photo)

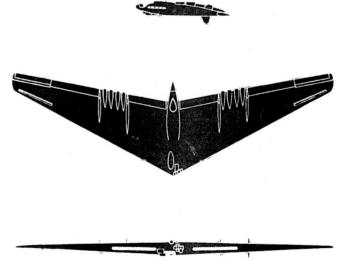

This official drawing of the YB-49 jet-powered Flying Wing shows the amazing thinness of the fuselage. Note that the wing is absolutely straight with no upward cant. Also note that the design omits the large twin vertical tails carried by the B-35 design, using four smaller versions instead. (USAF Photo)

The First Flying Wings 17

Four small vertical tabs outline the sets of jet engines on the trailing edge of the YB-49, shown here under construction at the Northrop production facility. (USAF Photo)

Looking at the head-on view of the B-49 design, it's easy to understand why the design had a certain amount of stealth capability even though it wasn't completely understood why at the time. (USAF Photo)

With its fairly tall landing gear riggers, the B-49 appears a bit ungainly on the ground. The small canopy bubble near the apex of the wing is visible from this ground photo. (USAF Photo)

The major part of this story, though, was the plane itself, which was, quite simply, awesome! But once you got through the mystery of its Flying Wing design, there were considerable similarities to "regular" aircraft, even though one would never have made that assumption gazing at the strange-looking, new plane.

The long, cantilevered wing would not have looked that much out of place installed on a bomber with a standard fuselage. But what was different from the straight wing time period was that they were swept back. The wings were coated with aluminum and were constructed in a single piece.

The size was such to allow an amazing crew size of 15, with the crew compartment being fully pressurized. Also, later

The smoothness of the top of the B-49 wing is visible from this topside photo angle. Notice the low fins that traverse across the top of the wing, starting at each of the small vertical rear-mounted fins. (USAF Photo)

This B-49 side view provides a better look at the small vertical fins which protrude both above and below the wing surface. (USAF Photo)

Up close and personal to the rear vertical fins shows the trailing edge flaps that butt up directly up to the outside of the fins. (USAF Photo)

The YB-49 prototype serves as the background for a band concert featuring the Air Force Band. (USAF Photo)

variants of the model would feature a rest area, which included beds for six of the crewman, along with a small galley. One has to compare that with the modern B-2 with a crew of only two, but with capabilities that surpass the B-35 by eons. Still, though, the similarities of the B-35 and B-2 are uncanny.

Propulsion problems were the biggest area of concern during the B-35 program. The four engines were each equipped with conventional four-blade propellers hooked to gearboxes. The complexity of the system would cause the company to later switch to single propellers on each engine. Those engines were, by the way, giant Pratt & Whitney R-4360 engines.

With all its technology and innovation, the possibilities of the B-35 were certainly promising, with plans on production for as many as two hundred of the model. But the fact was that the first flight test of an XB-35 would not take place until after the war, in June 1946.

That first flight lasted 45 minutes and was deemed as "Satisfactory and trouble-free" by the test pilots. Later versions of the plane would carry the YKB-35 designation. Shortly thereafter, though, the plane would experience gear box malfunctions and propeller control difficulties, which resulted in it being grounded in September of the same year as its first flight.

But there were other problems looming on the horizon for this machine, problems which would eventually cause its demise. One of the main problems came from the fact that the propeller-driven bomber could not meet the performance requirements. And, after all, this was the beginning of the jet era, and the Air Force was looking for more.

Also, the Flying Wing design was not as stable as the conventional design, thus causing its stock to falter even more. In all, there were 15 B-35s built; two were the X version, while all the others were the later Y models. A number of the planes would be diverted to a follow-on program, but a majority of the completed planes were scrapped.

The First Flying Wings 19

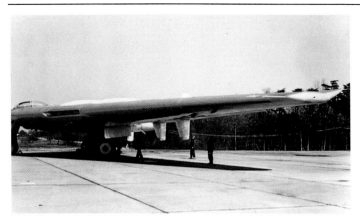
Considered as a highly-secret development at the time, the YB-49 sits on the tarmac at Andrews Air Force Base, Maryland. Note that there are a number of Military Police in position. (USAF Photo)

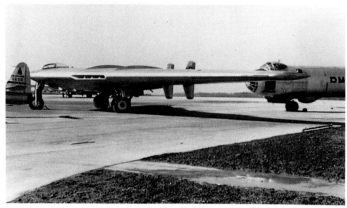
A comparison with the standard bomber design, as typified by the B-36 shown at the right, seem worlds apart with the B-49 Flying Wing bomber. (USAF Photo)

The thinking on strategic aircraft was turning a different direction. If the jet engine was such an improvement, why not implant a number of them into a B-35-type airframe? And that is exactly what happened with the formation of the B-49 program.

Note that the -49 designation was two later than the B-47, a six-engine jet bomber that would be built in huge numbers for the Strategic Air Command.

The B-49 program actually found its roots during the late World War II time period when a study was undertaken to determine the possibility of converting the prop-driven B-35 to jet power. In many ways, it was more of a continuation of the B-35 program, even though the former had been canceled.

The first B-49, a Y version, was actually derived from a conversion of the two XB-35 prototypes. The completion date was slipped more than a year to mid-1947, much of the time required to fit the plane with new stabilizing fins, replacing the stabilizing effect that the prop engines had provided.

The number of engines on the B-49 design doubled to a total of eight 4,000 pound thrust J35 engines. Initially, the design had called for only six. In addition, the new plane sported four large wing fences, and a redesigned leading edge ahead of and between each pair of fences. That configuration provided a low drag intake for each of the two sets of four engines. Beyond that, though, the two Flying Wings were very similar.

Looking more like an aircraft of the 1980s or later, the B-49 was a magnificent-looking bomber in flight. Hard to believe that it was a 1940s design. (USAF Photo)

20 Northrop-Grumman B-2 Spirit

Chapter 3:
B-2 Designs and Development

Since the B-52 had been designed and deployed in the mid-1950s, a number of different bomber concepts were investigated and discarded. The only concept that made it to production was the long-delayed North American B-1, an aircraft which has had a checkered career to date.

During the late 1970s, though, it was decided that the next bomber should have a low-observable capability, and a study was undertaken in 1981 by the two prime contractors, Lockheed and Northrop, who had the greatest experience in the "black art."

Of course, the Northrop experience came from its earlier Flying Wing designs, which unexpectedly had some low observable capabilities. The Lockheed experience was much nearer in time with the F-117 stealth fighter, which showed its tremendous capabilities in Desert Storm. The smart money was certainly on Lockheed to win the contract.

But it did not work out that way, as Northrop came out the winner to design and develop the so-called Advanced Technology Bomber, the early name for the B-2. The engineering proposal came from analysis from a pair of teams that independently worked on the solution. One team was considering low observables as the prime objective, while the second team was looking at improved aerodynamics.

Then came the amazing results from both groups who basically came up with the same Flying Wing design to fulfill both requirements. And also, even Lockheed concluded that the decades old Flying Wing design of Jack Northrop best fulfilled their design goals, too. The fact that three indepen-

The B-1B bomber preceded the B-2 and presented a completely different image from the B-2. The stealth capability of the B-1B greatly exceeded that of the B-52, with the B-2 exceeding the B-1B by a similar margin. (USAF Photo)

The two YB-49 test vehicles were test flown extensively, with the second plane crashing shortly after being accepted by the Air Force. It was reported that the plane was tumbling uncontrollably before impact. The crash brought back the long-standing worries about the stability of the Flying Wing design.

In the late 1940s, an evaluation was made comparing the B-49 and the B-29. The problems revealed during the -49's flight test program relegated it to a losing position and pointed to its eventual demise. Another crash in March of 1950 secured its fate.

But, like the B-35 which evolved into the B-49, the B-49 was not quite dead, as there would be a follow-on RB-49 variant, a photo reconnaissance version.

Basically, the model would be a stripped-out YB-49 that would have only a bombardment capability, along with extensive photographic equipment. A number of different versions of the model were planned, including one with eight General Electric J47 engines, another with six General Electric J40 powerplants, and a final, ultimate configuration combining both turboprop and jet engines.

But, like all its predecessors, this Flying Wing would not make it either. One version, though, would be constructed, although the final propulsion suite would include six engines instead of the eight used on earlier B-49s. And the interesting aspect of those engines was that four were mounted internally, while the other pair operated on the outside of the wing. An advantage of the configuration was that the wing could carry considerably more fuel, allowing an impressive increase in range.

This amazing overhead photo shows the fleet of YB-35s and YB-49s aligned together. Unfortunately, this would be the sight of their demise since they were disassembled at this location at the Northrop facility. (Northrop Grumman Photo)

In retrospect, the Northrop Flying Wings might just have been a little ahead of their time from a technological point-of-view. Then, too, there was that strange shape that was hard for many to accept.

Well, things were no different over three decades later when that same strange shape evolved with the B-2 design. Officials today still wonder about the plane, but it has gone on to operational service, something those early Flying Wings were never able to accomplish.

Maintenance is performed on the add-on jet pods of the modified B-49. Note the minimal clearance of the pods off the ground. (USAF Photo)

The YRB-49 prototype is shown in flight with the add-on pods clearly visible. Without doubt, the addition of the pods greatly decreased the stealth capabilities of the design. (USAF Photo)

Look real quickly at the take-off of this Flying Wing bomber and there might be a first impression that this is a B-49. But look again, it's the modern B-2 showing the huge similarities between the five-decades-apart designs. (Northrop Grumman Photo)

The addition of the lower jet pods provided a completely different look for the B-49 design. The pods were very similar to the pods that hung under the B-47. (USAF Photo)

The First Flying Wings

The key aeronautical term of the 1990s has been 'Stealth', a virtue carried significantly by the B-2 Spirit. But the B-2 wasn't the first to be designed from scratch with a stealth capability. The F-117 was also developed with that criteria, but the 'invisible' aspects of the model were achieved in a different manner. (USAF Photo)

Many programs during the 1990s were vying for money with each Service feeling that their program was the most important. From a Navy standpoint, the F-18E/F (shown here) deserved its fair share. (McDonnell Douglas Photo)

dent studies came up with the same basic design had to be gratifying to the Air Force who had to figure that it had the right design.

The strange shape fulfilled a number of different design criteria, including the lowest radar return head-on and the best lift-over-drag ratio for the best fuel efficiency in long range flight.

And even though the B-2 in its final configuration would look suspiciously like its earlier Flying Wing brothers, modern technology would be available to make it into an entirely different beast. The fast-growing area of flight controls, plus the great advances in flight structures, especially in the use of composite materials, would play heavily in the development of possibly the last manned bomber.

Like all of the major Air Force aeronautical systems, the program would be managed in great secrecy by the Air Force Aeronautical Systems Division (now Aeronautical Systems Center) at Wright Patterson Air Force Base, Ohio. Rumors circulated for years about the program, but its capabilities—and most importantly, its amazing shape—remained in the dark until the first details of its shape were released in the late 1980s.

B-2 Designs and Development 25

The C-17 transport is another expensive program of the 1990s struggling for funds. Many times, it was mentioned in the same breath with the B-2. (USAF Photo)

The mission goal of the B-2 in the beginning was to design a bomber capable of evading the extensive Soviet radar capabilities, where there had been huge amounts of money spent by the Russians.

The original B-2 was also conceived as a high-altitude bomber, a regime that obviously produced contraits which provided evidence of the plane's presence from the crudest of sensors, the human eye. The exact solution to the problem was never revealed, but B-2 officials indicated that the problem had been solved.

Of course, the stealth capabilities of the model had possibly the highest priority in the plane's design. It would appear that the design certainly accomplished that goal, as the final B-2 configuration was close to its 0.5 square foot radar cross section area. That minimal cross section compared to the 10 square feet area of the B-1B and the monumental

The Air Force, during the same period as the B-2 program, was developing possibly the most expensive aircraft program in history. The F-22, shown here in the foreground, competed head-on-head with the B-2 for funds. (USAF Photo)

When the B-2 was initially rolled out, this particular plane being Air Vehicle 1(AV-1) was the first completed model and was used in the flight test program. Later, the plane would be modified and converted into an operational configuration. (USAF Photo)

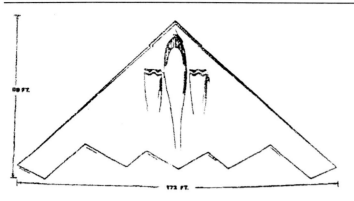

In the late 1980s, the Air Force decided that it was finally necessary to divulge the shape and size of the B-2, which to that date had been kept under wraps. The sketch, as is very clear, only showed the minimal details of the dramatically-new bomber system. (USAF Sketch)

1,100 square feet of the venerable 1950s-vintage B-52 bomber.

The radar effectiveness of the Flying Wing design, though, really should not have been that surprising to engineers, as the YB-49 version decades earlier had shown the capability. Radar operators at the time noted that the YB-49 did not appear on radar screens until it was directly overhead!

In 1984, a reassessment of the expected threats the B-2 would face was accomplished at the monumental cost of about $1 billion dollars. The resulting redesign involved changes in the carry-through structure where the wing halves met in the center of the aircraft to aid in withstanding the increased loads encountered in low-level flight.

The redesign was funded by the Air Force, and fortunately was accomplished before the plane was in production, where the change would have been many times more expensive. Even so, the redesign resulted in moving the projected first flight date to 1989 from an earlier projected 1987 date.

Initially, the crew for the B-2 was considered at three members, but during the design process a third member would not be added. But probably the most unsettling influence on the design program for the B-2 was the unending assault on the plane because of its unbelievable cost and the fact that many considered the plane unnecessary in the world of the 1990s.

The cost was, of course, very contingent on the number of models that would be produced. And that figure was also very fluid, varying from the initial 132 production estimate to the final minimal 21 built. At press time, there was still some deliberation about the possible construction of nine additional planes, but that possibility seems practically nil.

The cost of the B-2 during these design and development days had a serious effect on the overall defense budget. In fact, prompted by reports of the increasing cost of the program, an independent cost review was ordered to be un-

The first showing of the B-2 was well orchestrated with the rear of the plane facing away from the assembled group. (USAF Photo)

The head-on appearance of the B-2 bomber was unique in all the history of aircraft. This angle shows the protruding cockpit and the twin engine intakes which rest close to either side of the cockpit. (USAF Photo)

dertaken by the U.S. Comptroller General. It was a time when other major aerospace programs were clamoring for dollars, programs such as the C-17, F/A-18 fighter, USAF Trainer Program, and others.

In addition to its budget implications, the cost of the B-2 also had an effect on the problem-plagued B-1 program in the late 1980s. The thinking was that if perceived or real problems with the B-2 design caused the plane to be terminated, the B-1 would have to fulfill its penetration missions. B-2 problems also brought forth the arguments of advocates for completely different types of strategic offensive weapons, such as mobile ballistic missile systems.

Through the "black" design and development program, details of the actual design were also kept amazingly in the dark. But in 1988, it was decided that the first roll-out would occur in November, showing the Flying Wing configuration to the world.

The exposure was the first look at the design, and varied from the manner the F-117 Stealth Fighter configuration had been earlier introduced. That highly-unorthodox design first saw the light of day via a drawing released by the Air Force. With the B-2, though, it was felt that actually showing the unique craft was necessary.

The Air Force explained that the public showing of the plane really indicated the transition into the flight test program, but it was also certainly a publicity opportunity. But even though the plane was open to public exposure, there was still an air of secrecy about the details of the plane's external shape and backside details. To that end, the plane was parked close to the hangar doors with detailed photography of that rear area prohibited.

In the early 1990s, the decision was made to give the bomber a conventional capability. By the mid-1990s, that capability had been designed and successfully tested. Use of such an expensive system to carry "iron bombs" brought forth raised eyebrows, but when you think about it, there was nothing else available to do the job with the ancient B-52s and still-to-be-proven B-1s.

It was fitting that during the B-2 design and development period, Jack Northrop—who pioneered the Flying Wing concept in the 1920s—was given a special briefing and viewing of the B-2 before his death in 1981.

Chapter 4: B-2 Parts and Pieces

When a "normal" bomber configuration is described, such dimensions as the length of the fuselage, vertical tail height, and whether the plane is a high or low wing configuration are tabulated.

Such dimensions, though, are just not prudent when describing the B-2. When the dimensions of the wing are described, i.e. the span and length of the wing, the complete aircraft is being described. After all, this is a Flying Wing!

That wingspan, by the way, measures out at 172 feet (which is 13 feet shorter than the B-52). There is one interesting comparison when you lay the B-2 over the B-52 and discover that the wing angle is almost identical between the pair. What is also amazing is to learn that its about the same length as the F-15 fighter. But seriously, with their two immensely different configurations, that final comparison might be more of an apples and oranges situation.

Looking like a giant manta ray sliding through the sky, the B-2 introduced a configuration that had everybody's eyes blinking the first time it was revealed. (USAF Photo)

The sun highlights the only changes in the smooth flow of the B-2's upper wing surface. Those raised areas include the intakes for the bomber's four engines and for the crew compartment that is centered on the plane's wing surface. (USAF Photo)

To better understand the size of the B-2, you could line up four F-117 stealth fighters, wingtip to wingtip, and the B-2 span could encompass all of them. Lay it on a football field and it would stretch from the home end zone to the visitor's 40 yard line! In all, there is a monumental 5,140 square feet of wing area.

The stealth bomber is over ten times longer than it is tall (at only 17 feet), and has a wing width at its widest point of 68 feet. The B-2 has been described as having a "sinister" look about it, and it definitely does. The best description might be that of an attacking bat.

It is a heavyweight, weighing in at 153,700 pounds empty and 336,500 pounds fully loaded. Its payload capability is about twenty tons. Flying a high-altitude mission, the B-2 has an operational range of about 6,000 nautical miles. The bomber also has a 50,000 foot altitude capability in a terrain-following mode.

The configuration can best be described as a huge Flying Wing with some 5,140 square feet of surface area. It features smooth surface contours, unbroken outer surface lines, and extensive use of composite materials. Its quartet of powerful engines are recessed with top-of-wing inlets and exhaust. Every aspect of the plane was designed from the beginning to minimize signatures in radar, infrared, acoustic, and electromagnetic emissions.

Over 200 computers control and integrate functions on the B-2. These are grouped into three functional areas: avionics, flight controls, and aircraft subsystems. B-2 baseline sensor suites include state-of-the-art threat warning systems that are designed to accommodate improvements in sensors and avionics as they become available.

The B-2 is also capable of handling a number of weapons, both conventional and nuclear, in totally enclosed weapons bays. These characteristics were planned to assure a 30-year operational capability for the bomber.

On the ground, the B-2 presents an extremely low silhouette with a look more like a marine vehicle than one that would take to the air. (Phil Kunz Photo)

USAF/NORTHROP B-2 BOMBER

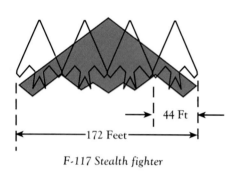

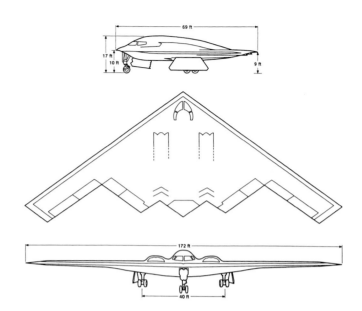

This three-view line drawing displays the exact shape and size of the B-2 Spirit Bomber. (USAF Drawing)

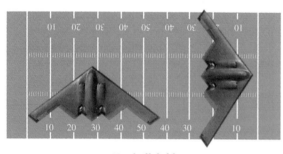

A size comparison of the B-2 bomber is shown, first being overlaid with the longtime B-52 bomber which has slightly longer wingspan than the B-2. Also note the fact that four F-117 fighters could be lined up side-by-side within the B-2 wingspan. Finally, the B-2 size is compared with a football field stretching from one end zone, across the center of the field and on to the other 40 yard line. (USAF Drawing)

In flight, the B-2 presented an out-of-this-world appearance showing that this was indeed a different type of animal. (Phil Kunz Photo)

It goes without saying that the B-2 is probably the most complicated bomber ever built. But the same can't be said for the maintenance of the machine. Maintenance crews have indicated that keeping the B-2 up to snuff is certainly a lot easier than recent-vintage bombers.

The system, for example, has the capability of precisely identifying the location of the problem and exactly what is wrong with it. Every electrical and mechanical part on the bird has the capability of sending a message to the inflight computer when it fails. And the magic continues with the flight-control computer being able to diagnose its own problems.

Let's take a closer look at this fabulous flying machine:

The Wing Surface

It is a strange situation when discussing the configuration of the B-2. When discussions of the wing are made, it is really a

32 **Northrop-Grumman B-2 Spirit**

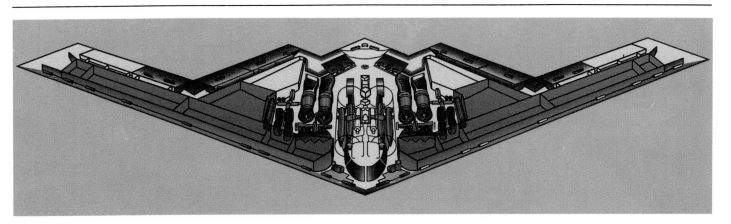

This cutaway drawing clearing shows the innards of the Spirit bomber. The major components of the plane, namely the crew compartment and four engines, are lumped together on the centerline of the wing. Also visible is the construction technique of the complete wing structure. (USAF Sketch)

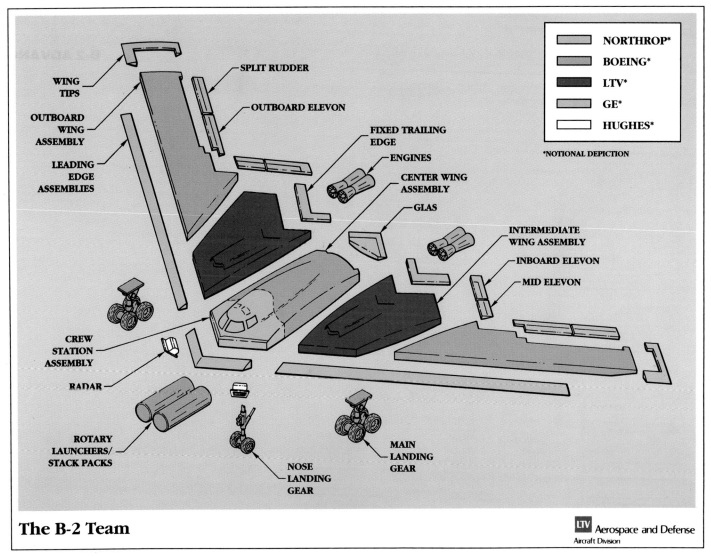

This early drawing shows the major components of the B-2 bomber with the parts color-coded with the contractors that built them. (LTV Photo)

B-2 Parts and Pieces 33

No worry about having to think about clearance of the rear fuselage of the B-2 while maneuvering on the ground. The reason is simple as there is no rear fuselage on this machine. (USAF Photo)

discussion of the plane itself, since the wing is the plane. One of the most visible aspects of this bat-like aerodynamic surface is the cleanness of the wide sweeping surface. No vertical appendages topside for flight control, and no pylons on the underside for fuel tanks or weapons.

With a wing area more than twice that of the B-1, there is what is known as a flat-lift curve slope design with elevons used for ride control. Flight controls are driven by hydraulic actuators. The control system for the bomber is a highly-redundant fly-by-wire system.

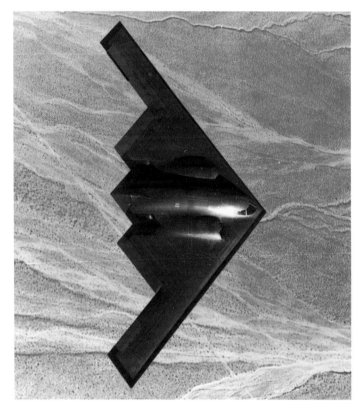

The sawtooth pattern of the wing trailing edge is certainly evident in this overhead photo. (USAF Photo)

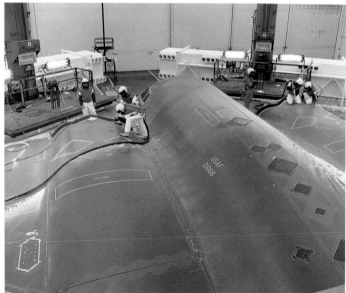

The flowing lines of the B-2's upper wing surface with the crew compartment and engine bulges clearly visible from this perspective. (USAF Photo)

34　Northrop-Grumman B-2 Spirit

As this B-2 Spirit moves in for a refueling operation, the sharpness and the wing's leading edge can be seen. Also note the pointed upper surface of the top of the engine cowlings. (USAF Photo)

The use of extremely strong composite structural members make the B-2 a very rigid aircraft, unlike earlier bombers (such as the B-52), which had a large wingtip up-and-down deflection during flight. Flight testing has shown that normal wingtip deflection is only about three feet. That is a huge advantage when stress on the fuel tanks is considered, which results in almost a total elimination of leakage problems.

The B-2's eight fuel tanks are obviously all carried internally. Drag minimization was definitely a design goal here. Even so, the plane does not possess supersonic capabilities, even though it has a Mach 3 look about its design.

Viewing the upper surface of the wing, the smoothness is extremely evident with the expected pitot tubes, sensors, and other protuberances just not there.

This inflight B-2 photo shows deployment of the end-of-the-wing drag rudders in their fully extended position. (USAF Photo)

This B-2 is in a hard bank to the left clearly revealing its trailing edge configuration. (USAF Photo)

The design is emphasized by the bulbous cockpit, which features wide, wrap-around windshields and protrudes from the smooth wing upper surface. The point of the cockpit is actually the front of the aircraft.

Two smaller paralleling protuberances flank the cockpit, each of which houses a pair of General Electric F118-GE-100 engines. The intake locations, though, are set back from the point of the cockpit.

The leading edges of the engine intakes are scalloped with forward-tilted leading edges. The engines are buried deep in the nacelles, which allows the engine compressor faces to remain out of reach of radar. The engine exhaust is ducted over the carbon-carbon surfaces on the wing's upper surface and trailing edges.

Other visible characteristics of the top wing are four small auxiliary engine inlet doors, which open outward on the up-

Details of the trailing edge control surfaces. Also visible are the large landing gear doors. (Phil Kunz Photo)

A deployed flap on the trailing edge of the B-2 bomber wing. (Phil Kunz Photo)

36 Northrop-Grumman B-2 Spirit

The front bottom of the B-2 appears to completely open up with the front landing gear is deployed. Note the sawtooth pattern on the top and bottom of the landing gear door which aids to its stealth capabilities. (USAF Photo)

per side of each engine nacelle. The doors, when open, also serve to provide cooling outside air to the engine bays while the aircraft is on the ground.

A sharp line culminates the joining of the lower and upper wing surfaces. The angle of those two lines establishes the sweep angle of the plane, an angle which is only slightly greater than the XB-49, with its span being practically identical to the earlier Flying Wing.

The Flying Wing design of the B-2 is basically an unstable shape, requiring the flight control system to perform to perfection one hundred percent of the time. The sophisticated fly-by-wire system, though, has proved to be up to the tough job.

Even with its impressive range, the B-2 is still capable of mid-air refueling, providing it with unlimited range capabilities. The refueling probe is contained in the top wing surface aft of the center cockpit location.

The underside of the wing is relatively flat from the front tip to the mid-span point, where it then slopes downward, increasing the cross-sectional area of the wing in this area. This increase was necessitated to contain the plane's weapon systems, all of which are housed internally.

The rear of the wing sports the now-well-known sawtooth shaping. There are five "points" which define the outer ex-

Describing the B-2 as having a needle nose can certainly be verified when viewing the plane from this angle. The sawtooth pattern is visible atop and on the bottom of the engine intake. (USAF Photo)

The front bottom of the B-2 appears to completely open up with the front landing gear is deployed. Note the sawtooth pattern on the top and bottom of the landing gear door which aids to its stealth capabilities. (USAF Photo)

tremities of the rear of the plane. The middle three points are centered on the centerline of the cockpit with the center point lying exactly on that line. The outer pair of points extend out further than the inboard three points.

Directly behind the cockpit area on the rear of the wing is a flap-like device which is hinged at the trailing edge to allow it to be deflected vertically.

Recall again that the B-2 has no vertical tails, but the trailing edges of the wing carry long control surfaces. Deflection of these surfaces effects roll, pitch, and yaw responses for the aircraft.

Also contained on the lower wing surface are the three main landing gear locations. The twin-bogey main gear is located near the center of the wing underside some 40 feet apart. Manufactured by Boeing, the main gear is very similar to the company's 767 commercial transport unit.

The nose gear is located near the forward portion of the plane near the point of the wing leading edge. The nose wheel gear is divided into two parts, while the main gear doors are large triangle slabs located on the outside of the gears in the down position. It has been noted that because of the size of

The main twin-bogey landing gear for the Spirit wasn't a technology breakthrough with use of an existing commercial version being employed for this application. (Phil Kunz Photo)

The landing gear position is shown just after lift-off and rotation. There's a lot of weight to be rotated into the fuselage once the ground is cleared. (Phil Kunz Photo)

these doors, when the gears are deployed at slow speeds, the B-2 responds somewhat like a normal aircraft.

Both main gear units are four-wheel, tandem-dual trucks, which are attached to a single strut. The front gear is a single-strut unit with two wheels. A pair of landing lights are mounted on the left side of the gear strut.

The brakes within those landing gear units have anti-lock capabilities. And they can think, too! When a brake is about to lock up, the pressure is released and the load is shifted to the other units, thus keeping the plane from skidding on landing, and destroying a tire.

The wing's leading edge, by the way, is completely smooth thanks to a process designed by Northrop to internally connect the sections together.

Contained in the lower leading edge of the wing are two antennae of the plane's sophisticated LPI intercept radar system.

Cockpit

The cockpit for the two-man crew is neat and uncluttered, with the pilot sitting in the left seat. At one time, there were considerations for making the crew to number three, but those considerations have been shelved.

Crew ejection is accomplished by an upward firing system for each seat. Entry into the compartment is accomplished through a hatch at the aft end of the nose gear wheel well.

The engine throttles are located to the left of each crew member. With his position in the right seat, the mission commander has easy access to the cursor controller and data entry panel. Compared to earlier bomber aircraft, the number of dials and panels appears considerably fewer with the B-2.

There are dual flight controls with a fighter-style stick, which is very similar to that of the B-1. Just to the right of the Mission Commander is the so-called mission recorder, where a tape with the mission data is loaded into the aircraft's computers. The bomber has the capability to be retasked while in flight. Contained in both crew positions are four Multipurpose

The B-2's two-man crew resides in a compact and beautifully designed cockpit. Obviously, the dials and instruments are state-of-the-art. (USAF Photo)

Display Units (MDUs), which provide a myriad of information including the status of the fuel condition.

An interesting difference with the cockpit display from earlier USAF bombers is the lack of knobs associated with the turning on of systems. Those functions are performed electronically by on-board computers. The bezel buttons and multipurpose display panels also aid in the activation of aircraft systems.

The reduction to the two-man crew halves the number required for the B-1 and the final versions of the B-52.

Controlled from the cockpit is the high-tech Gust Alleviation System (GLAS), which is physically located in the rear of the aircraft. When the B-2 is in flight conditions, the GLAS is retracted into the fuselage and functional in a different role as an automatic pitch axis trimming surface. The cockpit also carries cathode ray tube displays and conventional control sticks.

Stealth Characteristics

Of course, one of the prime reasons for the B-2's unique shape is to achieve stealth characteristics. Although the plane did not evolve into the F-117-type configuration, there are a number of stealth-producing capabilities derived from its Flying Wing configuration.

First, the sweeping curves and streamlined shape are effective in diffusing and deflecting radar beams. That effect is implemented by the fact that the entire aircraft is coated with honeycomb ferrite-based material, which absorbs radar beams. The stealth considerations were carried internally where many internal structural elements are fabricated of layered carbon-fiber composites, which are less visible to radar

Up close and personal with the B-2 cockpit. Pilots reveal that forward vision is limited with the runway directly in front of the plane not in view. (Phil Kunz Photo)

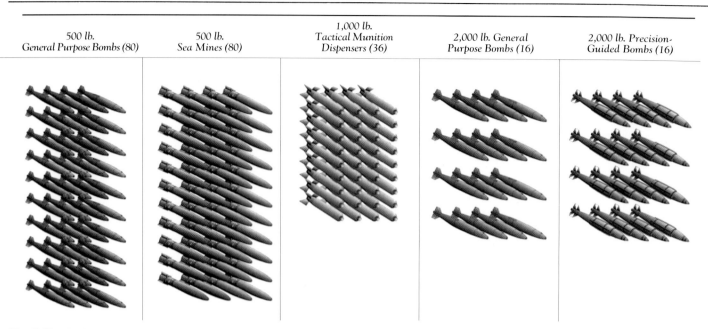

Flexibility is the key word when describing the different weapon suites that are available with the B-2. Here are five different weapon arrangements with General Purpose bombs, Sea Mines, Tactical Munition Dispensers, and Precision Guided Bombs. (USAF Drawing)

than metal. With engine exhaust heat tending to generate a significant radar signature, the engine exhaust gases are mixed with cool air and exited on top of the wing, away from ground based infrared radars.

The RAM (Radar Absorbing Material) that coats the external covering provides only about ten percent of the stealth characteristics of the plane, with the shaping of the plane being more important. Unfortunately, that high-tech covering brought about some problems during the plane's early operational days. The Air Force has since developed a technique to protect the RAM material.

The B-2 also has the advantage that it does not have any sharp edges or vertical surfaces exposed, making it difficult for radar beams to bounce off and return to the transmitter.

B-83 bombs are loaded during testing at Edwards Air Force Base, California. (Northrop Photo)

The ground support equipment for the B-2 is shown here in a loading exercise for an early B-2. Note the weapon bay doors which are in the open position. (Northrop Photo)

Of course, with any new high-tech program, there are skeptics as to its real capability. That was sure the case with the B-2, which endured much more than its share concerning its stealth capability.

In 1989, for example, the existence of a new Australian-developed radar that could possibly defeat the B-2's stealth protection was announced. And even more recently, the stealth capability of the B-2 was really questioned when it was reported that the stealth capabilities of the plane were greatly reduced when the plane was operating in rain.

The amount of maintenance required on the B-2 has also been criticized, along with the fact that use of the low-observable equipment effects the mission capability aspects of the plane. Reportedly, the B-2 demonstrated a 22 percent mission capable rate with the low-observable equipment engaged, while it is raised to 69 percent with the equipment not engaged.

Maintenance on this complicated aircraft was considered from the earliest days of its design. The plane was designed to be maintained by Air Force technicians with standard education and skill specialties.

But the plane has survived and continues to soldier on through all the criticism and bad publicity. The fact that the B-2 would enter the target area subsonically might surprisingly provide an enhancement of its non-detection capability. Even though the plane will not have the capability to approach the target supersonically, it will enter without the "supersonic footprint."

Interestingly, there is little aerodynamic effect on the performance of the B-1 because of the stealth design because the Flying Wing design has excellent flying capabilities. In order to achieve the desired radar cross-section, it was necessary to have precise conformance of external geometry and dimensions to the desired shape. The surface of the aircraft also possesses electrical continuity and smoothness. The dimensional controls of the aircraft are so tight that there is less than a half inch variance between the wingspans of any two aircraft.

Weapons

Again, for aerodynamic reasons, there are no external weapons mounts, and that situation will probably remain that way through the life of the aircraft. Having any external protuberances on the plane would greatly reduce the stealthiness of the bomber.

The selection of weapons (these by the way are all offensive weapons) is sizable, with nine conventional and two nuclear weapons. Like the B-1, the B-2 is the second straight bomber to have no guns—none! But that seems to be the current trend, as even the venerable and still operational B-52H recently had its rear stinger gun removed.

There are a pair of side-by-side weapons bays. The actual launching occurs from a Rotary Launcher Assembly (RLA), which is capable of mounting up to eight weapons in each bay, or two Conventional Bomb Assemblies in each weapons bays.

Loading of weapons onto the rotary launcher in the B-2's massive weapon's bay. (Northrop Photo)

It is extremely surprising to note the volume of the bomb bay considering the apparent thinness of the wing. The following is a listing of the weapons that constitute the B-2 weapons suite:

Conventional Weapons

Weapon	Class	Load	Weapon Type
MK-82	500	#80	General Purpose
M-117	750	#36	General Purpose
MK-84	2,200	#16	General Purpose
TSSAM	-	8	General Purpose
JDAM-I/III	2,000	#16	General Purpose
CBU-87/89/97	1,000	#36	Tactical Munition
M117	DST750	#36	Mine Destructor
MK-36	500	#80	Mine Destructor
MK-62	500	#80	Mine Magnetic Fuse

Nuclear Weapons

Weapon	Load	Weapon Type
B-61	16	Gravity Bomb
B-83	16	Gravity Bomb

In early 1997, B-2s were provided with a modification of the B-61 bomb. The externals of the weapon were changed with the elimination of the drag parachute and the addition of an aerodynamic fin, allowing for greater target-hit precision. There was also a steel nose added for greater penetration. The bomb's added velocity increases its penetration capability to three to six meters, making it very effective against deeply-buried targets.

The B-2 can carry up to 16 nuclear weapons or 20 smaller versions. Or, it can load up to 80 conventional 500-pound weapons, along with sea mines and other ordnance. That's up to 40,000 pounds of either nuclear or conventional ord-

The loading operation of a 4,700 pound GAM-113 weapon is completed and the bomb bay is ready to be buttoned up. (Northrop Photo)

The free-fall drop of the GAM-113 bomb from a B-2 during the flight test program at Edwards Air Force Base. (Northrop Photo)

A salvo of Mk-84 bombs cascade from this B-2 during the flight test program. Notice the bombs are falling in pairs as they emerge from the bomb bay, and as they continue to fall, they begin to separate from each other. (Northrop Photo)

The business end of the B-2's propulsion system is shown by the engine cowl that accepts the incoming air. The arrangement is the first of its kind in aviation history. (Phil Kunz Photo)

nance. It has been reported, though, that it is unlikely that the bomber would ever carry that full load in an operational environment. Understandably, it is felt that the real efficiency for the B-2 will be to carry a smaller number of the smart weapons.

The Global Positioning System-Aided Targeting System/GPS-Aided Munition of GATS/GAM was developed by Northrop and Hughes Aircraft to provide the Air Force with an interim all-weather, launch and leave capability until the Joint Direct Attack Munition (JDAM) was expected to be deployed in the 1999-2000 time period.

The Engines

A derivative of the F101 and F110 engines, the General Electric F118-GE-100 engine is capable of 19,000 pounds of thrust. Four of these engines power the B-2 to high subsonic speeds.

The engine has no afterburning capability, and employs new long chord fan technology with the same compressor and turbine used on the F110 engine, which results in higher thrust. The engine was fully qualified in 1987.

The F118 uses a three-stage fan with variable inlet guide vanes, a one-stage high-pressure turbine, and a two-stage low-pressure turbine. The engine measures 100.5 inches long with a diameter of 46.5 inches.

The super efficiency of the engines, plus the super-slick aerodynamics of the airframe design, all blend together to produce the range capabilities of this plane. The engine/aircraft design also provides the capability of operating on shorter runways—shorter than certain commercial airliners. It also points to the capability to operate on many civil airport runways, something that could be a huge advantage for emergency situations.

Since the powerplants are buried in the fuselage in order to hide their signature from radar, it makes them a little more difficult to service. In early experience with the plane, though, there have been no big problems.

B-2 Parts and Pieces 43

Chapter 5:
B-2 Production

B-2 production planning has definitely had a downward spiral through the years, really moving somewhat the opposite direction of the cost estimates for the plane.

From the very beginning, the number of aircraft to be built was very small (132), when compared with bombers of the past. But when you think of it, the B-1 really set the trend in this low-production situation with only 100 built.

Later the number 75 would appear, enough bombers to equip two full wings. Then, the monumental drop when it was decided in the mid-1990s that only 20 models would be built. Later, it was decided that an early flight test model would be converted to operational status, raising the number to 21.

But the story was not quite over. In June 1997, there was pressure from some quarters to continue production beyond the 21 limit. The House Defense Authorization Bill contained $331 million for production of an additional nine aircraft. However, in the time period, it appears that any additional production of B-2s is highly unlikely. The 1998 Defense Budget did include additional funds for the bomber, but none of it was appropriated for additional production. Instead, it called for additional improvements to the existing 21-plane fleet.

Surprisingly, the Air Force was not for the additional aircraft, saying that it had enough of the model to fulfill its mission requirements. Proponents for the program argued that its deep-strike capabilities and large payload justified its continued production in view of the fact that there was nothing else to replace it.

What it really came down to, though, was that there was worry that the money for the additional B-2s might effect the F-22 program. Also, even the other services figured it might effect their major programs, too. With the defense budget continuing to decrease, and every major program's cost continuing to spiral, something was going to have to give. Many argued that it should be the B-2, while others had the opposite opinion.

The uncertainty of the B-2's future also effected the production of the plane. Congress was constantly debating on the production numbers, and whether the plane should be produced at all. But when you really get down to it, most super-expensive military programs face the same uncertain situation as the costs continue their inevitable climb. In 1989, there was even a vote to completely kill the program, but that was defeated by a 279-144 vote.

It seemed that the ultimate disposition of the B-2 lay somewhere between total cancellation and full production. A majority of the opinion seemed to favor one or the other with no in-between. Well, as is known, the final solution was somewhere between with limited production. But whether you were a strong or non-advocate of the program, the "sticker shock" price tag of the Flying Wing caused skeptics in every quarter.

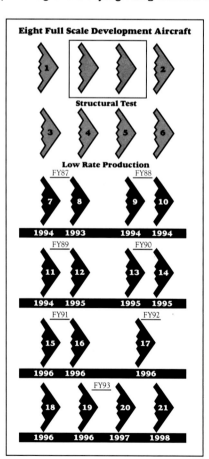

This production schedule shows the sequencing of the 21 B-2s that were produced. (Northrop Drawing)

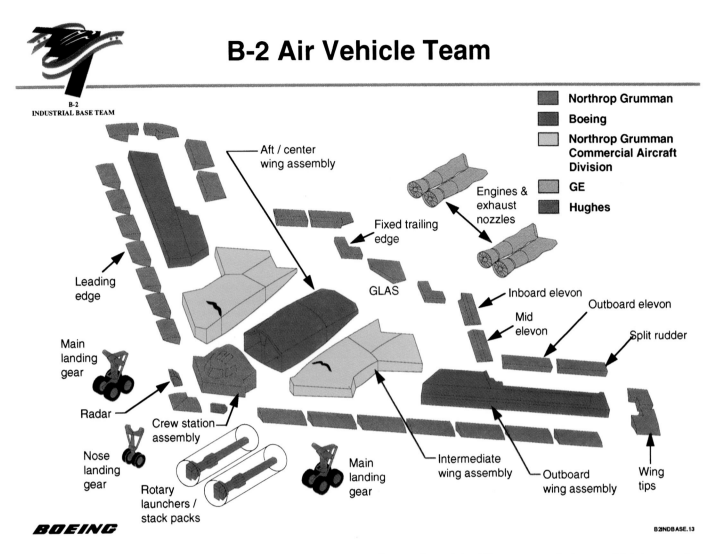

From a production point-of-view, the massive B-2 program was really a project of the US aerospace industry which is very clear when looking at the industrial base involved. (Boeing Drawing)

Fabricated from advanced composite materials, a large structural section of the B-2 is seen in production at Boeing. Seen in the background is the world's largest autoclave, a specialized oven which used heat and pressure to harden composite parts. (Boeing Photo)

Of course, production numbers of the B-2 would directly affect the production numbers of the other advanced programs, namely the Air Force F-22, the Navy's F/A-18E/F, and the still-to-fly Joint Strike Fighter (JSF).

Through the years, though, there were delays in production that slowed the actual data points from the initial projections. Even through all the turmoil, though, the first B-2 rolled out of the Northrop production facility in Palmdale, California, in November 1988. There was a sense of urgency to get that first plane before the world's eyes, with workers being diverted from planes number two and three to assure that number one was completed.

Northrop and the Air Force made the roll-out a national media event of the first order. Not surprisingly, there were a number of speeches listing the virtues of the radically new, and obviously the extremely expensive, bomber.

Air Force Chief of Staff Larry Welch indicated at the time, "The B-2 is a superb example of the inherent technological

Assembly progress on the B-2 outboard wing is shown in this January 1994 photo. At the time, the fuel system was nearing completion along with the installation of the electrical system. (Boeing Photo)

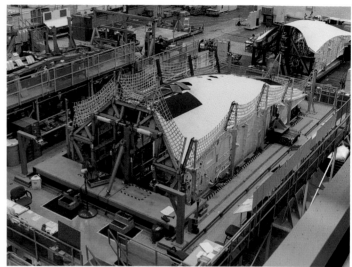

This photo shows the assembly progress of a B-2 Aft-Center Section. The ACS is often referred to as the nucleus of the B-2 since it is where almost all parts and systems for the aircraft come together. (Boeing Photo)

advantage of a free society. This aircraft combines all the best attributes of the penetrating bomber—long range, efficient cruise, heavy payload, all-altitude penetration capability, accurate delivery, reliability, and maintainability."

Even though the B-2's shaping was still considered a sensitive area, the roll-out was still carried out, but the plane was positioned in such a manner as to hide the details of the rear sawtooth design of the aircraft.

Scheduling Implications

B-2 low rate production began in 1987 with the final B-2 being completed in 1998. A number of rescheduling decisions made through the years stretched out the production program later than its original plans. There were also a number

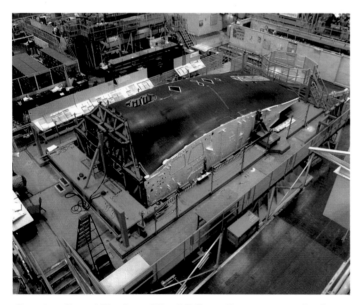

Construction at Boeing of the Aft Crew Compartment is shown by this photo. (Boeing Photo)

in Congress that wanted the B-2 to complete a successful flight test program before any further production was approved.

That famous first photo of the B-2 production line was an amazing sight, providing an idea of the complexity of possibly the final manned bomber to be produced. And even though the B-2 is the most complex aircraft ever built, the time for production of each plane decreased as the experience increased with each succeeding aircraft.

Block Modifications

Through the B-2 production phase, a number of different modifications have been made. But first, there needs to be an explanation on vehicle designation during the production process. First of all, all B-2s are identified with a "AV" prefix, for Air Vehicle, which is followed by the number of the vehicle from its production schedule. Then comes a slash (/), followed by the last two numbers of its year of manufacture, i.e. 88, 89, etc., and then four serial numbers.

The actual production of the B-2s was accomplished in Blocks, with AVs 6-17 being the initial Block 10 versions. AVs 17 through 19 were Block 20 planes, while the remainder were the final Block 30, which carried the most extensive modifications. The first full combat-ready Block 30 B-2, built from scratch, was completed in 1997.

The goal of the B-2 production program is to ultimately modify all the B-2s into the Mod 30 configuration, and that includes all of the initial flight test versions. To accomplish the Mod programs, the earlier version B-2s were returned to the production facility.

The Mod 10 versions had the capability to carry the B83 nuclear bombs or 16 Mk84 two thousand pound conventional munitions. The Mod 20 versions had increased weapon-carrying capability, with the addition of the B61 nuclear bomb

One of the key systems for the B-2 are its pair of rotary launchers, the last one shown being delivered by Boeing in 1995. The launchers can hold and release a variety of weapons. (Boeing Photo)

Wearing protective suits, Northrop technicians accomplish the paint process of a B-2 at the company production facility. (Northrop Photo)

Work on the upper wing surface is accomplished by Northrop technicians at the company's Plant 42 in Palmdale, California. (Northrop Photo)

and the GATS/GAM system. Up to 16 GAMs can be carried on the two rotary launchers.

The Mod 30 consists of full-capacity terrain-following and terrain-avoidance radar, improved navigation, all-weather operation and enhanced weaponry, including Joint Direct Attack Munition capability. It also features 19 radar modes, up from 11 on the Block 20 and from six on the Block 10. The Mod 30 in fact also addressed a number of problems in the earlier versions and came up with the appropriate fixes.

B-2 Subsystem Production Facilities

Boeing was one of the main B-2 contractors and fabricated the wing outboard sections, which are about 65 feet long, and the broad area behind the cockpit, an area which is called the Aft Center Section (ACS) and is about 50 feet in length. In addition, Boeing was also responsible for the landing gear, weapons delivery avionics, and fuel system, along with weapons bay doors and the metal keel. The components were ferried to the Northrop main assembly facility by USAF C-5 Galaxy transports.

Northrop fabricated the forward center section, including the cockpit with LTV, making the intermediate section of the wing, the portion that is over the wheel wells.

B-2 Final Assembly Production Facilities

Plant 42, the location of the B-2, is located some 35 miles north of Los Angeles, California. This is the same facility where the Space Shuttle, Lockheed L-1011, and Rockwell B-1 were also based.

The actual B-2 production took place at a large manufacturing hangar which is located in the so-called Site 4 of the location. It was just outside this building that the roll-out and

Mass production is definitely not the correct description of B-2 production. It's a slow, meticulous process as each plane received careful personal attention during the production build-up. (Northrop Photo)

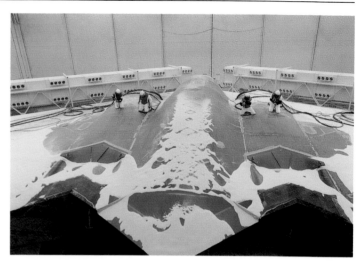

The paint removal process during a B-2 Mod program is shown here in progress. (Northrop Photo)

ceremony was held. The production facility is approximately 100,000 square feet in area, which included space for actual testing. As is well known, there was great secrecy with this program, and certainly, there was no sign on the building stating, "Home of the Air Force B-2 Bomber."

It has been reported that some forty thousand workers toiled on the B-2 program, something akin to the Apollo lunar program, in secret! As unconventional as the B-2 is, possibly even more unconventional was the way the bomber was manufactured. Use of a sophisticated three-dimensional data

Probably the most famous of all the B-2 released photos was this unbelievable shot of the production line with a half dozen of the bombers being built. (Northrop Photo)

48   Northrop-Grumman B-2 Spirit

A completed B-2 is prepared for delivery from the Northrop plant 42. (USAF Photo)

base enabled the prime contractor to eliminate the need for a number of prototypes. Since stealth capabilities were a prime goal, precise accuracy was attained with a computer-aided design system that allowed engineers to detail each component.

Also, the availability of the data base provided a significant production advantage, allowing the contractor to build production tooling without building the tools needed for prototype proofing.

The famous B-2 production line photos that were released provide an understanding of how the build-up was accomplished. Shown are a number of B-2s under construction, with the nearest B-2 being the furthest toward completion. The protective mats on the top of the wings is for protection of the composite skin.

The initial B-2s experienced some problems, but not unlike those encountered by all new aircraft systems. Some was caused by the vast new technology of the new bomber and the extremely exact tolerances dictated by the design.

Mod 30 work in progress on the near B-2 while production continues on another Spirit. (Northrop Photo)

The AV-1, the first B-2 produced, is shown undergoing final production tests at Building 42 before it is released to the Air Force. (Northrop Photo)

Chapter 6: Testing the B-2

It is probably not surprising to learn that America's most expensive military aircraft in history has also been involved in one of the country's most extensive test programs. The testing involved early simulation tests, ground tests on the first flight articles, and finally, a long flight test program.

Obviously, the goal was to reduce the chance of failure during the program, an occurrence which might have been devastating in view of the tenuous status of the program. The following is a discussion of the magnitude of the B-2 testing through the years.

Early Testing

During the research phase for the B-2 bomber, the unique shape of the B-2 was tested extensively in model form to assure its stealth capabilities. During the 1980s, more than 100,000 radar cross section images of B-2 models and components were tested.

In all, well over a half million hours of tests of all types were completed before the B-2's first flight. The testing involved wind tunnel, avionics, flight control, and computer systems analysis.

Alaska's northern lights give a B-2 flight test bomber a mystic look parked on the tarmac at Eielson Air Force Base during two week of cold-climate testing conducted in March 1995. (USAF Photo)

Don Vern
1st B-2 Test Pilot
Tom LeBeau
B-2 Test Pilot

USAF/NORTHROP GRUMMAN B-2

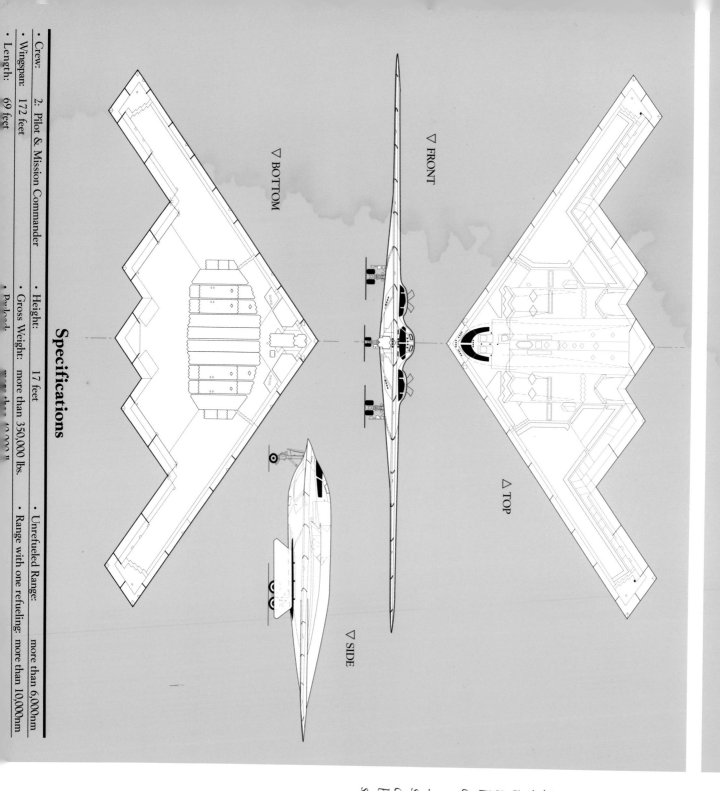

▽ FRONT
▽ TOP
▽ BOTTOM
▽ SIDE

Specifications

- Crew: 2; Pilot & Mission Commander
- Wingspan: 172 feet
- Length: 69 feet
- Height: 17 feet
- Gross Weight: more than 350,000 lbs.
- Payload: more than 40,000 lb
- Unrefueled Range: more than 6,000nm
- Range with one refueling: more than 10,000nm

THE B-2's ADVANTAGES

"So bombers are becoming a more important element of our forces as our overseas bases decrease, and we become more of a home-based expeditionary force. They have three great features - very long range, high payload, and the sense of immediacy; immediate impact on a crisis - so that, in my view, bombers can help forestall, can help deter a conventional conflict.

...the advantage of the B-2 will be to show any potential aggressor that he can be attacked within hours by B-2s, because of their stealth without any supporting aircraft."

Retired General John Michael Loh
Former Commander
Air Combat Command
December 10, 1993

USAF B-2 Spirit

Combat Proven in Operation Allied Force

"Nothing so well represented the Air Force capability to conduct global attack in the air war over Serbia as our B-2s delivering precision-guided munitions via 29-hour missions from Missouri to Yugoslavia and back."

"The combination of on-board systems and GPS guidance on the B-2 proved even more accurate than planners had expected. This meant the B-2 could precisely engage multiple targets per sortie, destroying a disproportionate share of total targets in some of the most heavily defended areas of the conflict."

General Marvin R. Esmond
USAF Deputy Chief of Staff
Testimony to the Senate
19 October 1999

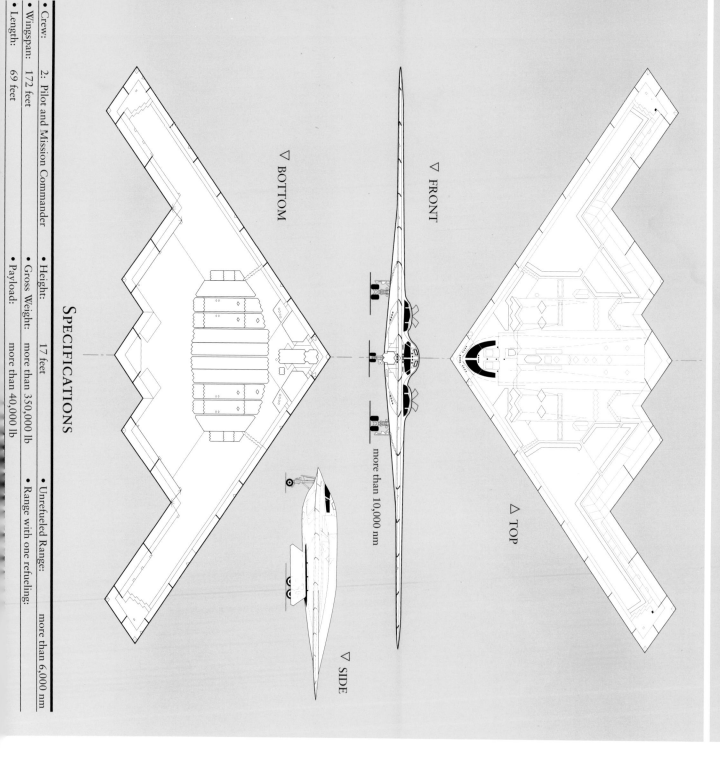

SPECIFICATIONS

- Crew: 2; Pilot and Mission Commander
- Wingspan: 172 feet
- Length: 69 feet
- Height: 17 feet
- Gross Weight: more than 350,000 lb
- Payload: more than 40,000 lb
- Unrefueled Range: more than 6,000 nm
- Range with one refueling: more than 10,000 nm

USAF B-2 Spirit

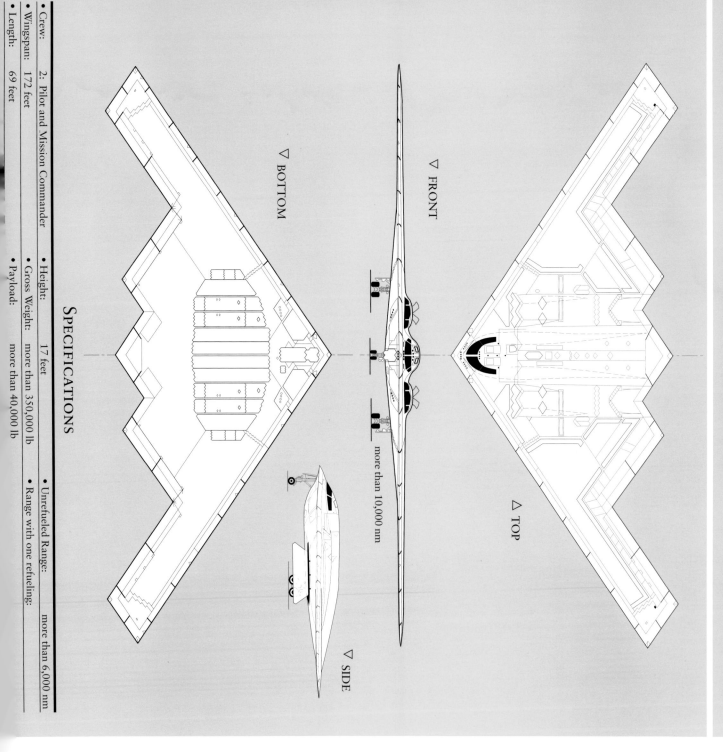

△ TOP
▽ FRONT
▽ BOTTOM
▽ SIDE

more than 10,000 nm

Specifications

- Crew: 2; Pilot and Mission Commander
- Wingspan: 172 feet
- Length: 69 feet
- Height: 17 feet
- Gross Weight: more than 350,000 lb
- Payload: more than 40,000 lb
- Unrefueled Range: more than 6,000 nm
- Range with one refueling: more than 10,000 nm

Combat Proven in Operation Allied Force

"Nothing so well represented the Air Force capability to conduct global attack in the air war over Serbia as our B-2s delivering precision-guided munitions via 29-hour missions from Missouri to Yugoslavia and back."

"The combination of on-board systems and GPS guidance on the B-2 proved even more accurate than planners had expected. This meant the B-2 could precisely engage multiple targets per sortie, destroying a disproportionate share of total targets in some of the most heavily defended areas of the conflict."

General Marvin R. Esmond
USAF Deputy Chief of Staff
Testimony to the Senate
19 October 1999

USAF/NORTHROP GRUMMAN B-2

THE B-2

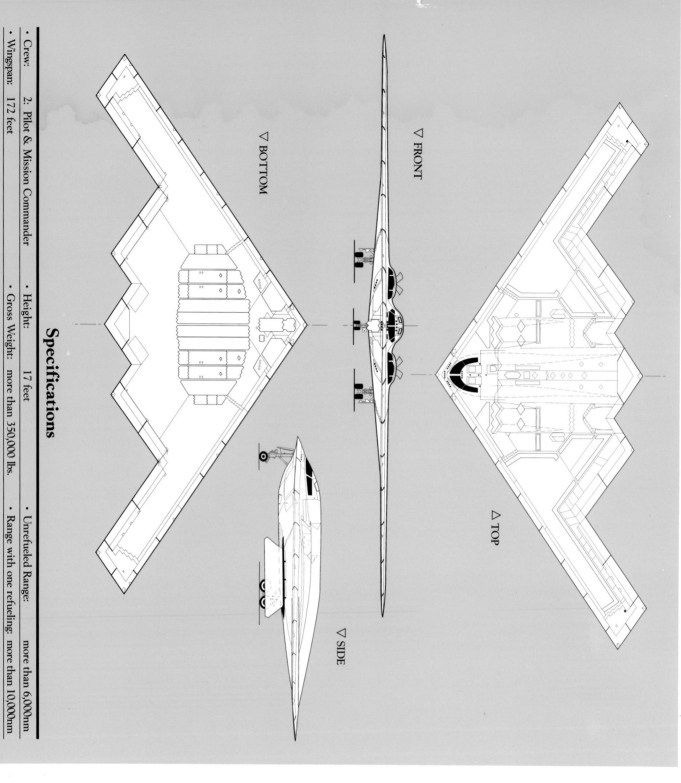

△ TOP
▽ FRONT
▽ BOTTOM
▽ SIDE

Specifications

- Crew: 2; Pilot & Mission Commander
- Wingspan: 172 feet
- Length: 69 feet
- Height: 17 feet
- Gross Weight: more than 350,000 lbs.
- Payload: more than 40,000 lbs.
- Unrefueled Range: more than 6,000nm
- Range with one refueling: more than 10,000nm

"The world is still an unsafe place and there is still ample need for a first-rate Air Force with first-rate, long-range bombers - fully capable of enforcing this nation's vital interest around the globe.

We must renew our commitment to remain strong around the world. And I believe that long-range bombers - especially B-2 bombers - are crucial to the Air Force's vision of "global reach, global power."

...the B-2 bomber, with its enormous range, stealth technology, and ability to deliver large amounts of conventional smart bombs, stands in the vanguard of our ability to respond swiftly and effectively to any threat of our national security.

Never has there been an aircraft more versatile and powerful than the B-2 bomber. Its superiority represents the nation's intention to maintain the best equipped military possible."

Senator Dianne Feinstein (CA)
"Spirit of California"
Naming Ceremony
March 31, 1994

Hot air ducting is in place to heat components of this B-2 during cold weather testing in 1995 at Eielson Air Force Base.

Long before the B-2's first test flight, the stealth bomber's complex avionics system was aloft. Using a specially-equipped C-135 testbed aircraft, the testing had occurred since 1986. The Hughes radar system was also tested on the testbed since 1987.

Ground Testing

Environmental testing was extremely important, and as such, cold and hot testing played an important role for the B-2. The initial testing took place at the McKinney Climatic Laboratory at Eglin Air Force Base, Florida, in the early 1990s. This huge facility has temperature tested all of the recent Air Force aircraft developments in recent decades.

Flight vehicle AV-5 was the first B-2 to go through the lab. For its testing environment, the bomber was frozen to -65 degrees and "baked" to 120 degrees F. But there was more, and the plane was also rained upon, snowed upon, scorched, and allowed to mold in extremely high humidity. The aircraft

No question from the way these technicians are bundled up that it is definitely cold at Eielson Air Force Base during B-2 cold weather testing. The B-2 achieved all planned goals during the testing in 1995. (USAF Photo)

Flight operations were a part of the cold weather verification process shown here with the test B-2 in flight in very Arctic conditions. (USAF Photo)

The maiden flight of the B-2 took place on July 17, 1989 at Palmdale, California. The flight would begin a long and vigorous flight test program. (Boeing Photo)

spent over one thousand hours in the climatic lab, which was more time than the planned flight test time for all the B-2s in the developmental flight test program.

But that was not the end of the environmental testing that would be endured by the AV-5 vehicle. Following the Eglin tests, the aircraft was involved in cold climatic testing at Eielson Air Force Base, Alaska. The Alaska testing concentrated on the functioning of the plane's Environmental Control System and Auxiliary Power Unit after being cold soaked overnight. The testing proved that both systems could operate properly in even these tough environmental conditions.

The only problems noted were frozen APU valves on several occasions. The test group, though, did not consider this phenomenon to be serious, as the valves could quickly be brought back on line by the application of heat. For the most basic of tests, the plane was left in the brutal Alaskan weather overnight with temperatures in the -20 degree range.

A B-2 test bomber is seen during a flight test over the Mojave Desert, California in 1994. It was one of many flights that proved that the B-2 was capable of meeting all its design goals. (Boeing Photo)

This dramatic photo shows a pair of B-2s on the ground, one of which is on the move, during flight test operations at Edwards Air Force Base, California. Note the near B-2's powerful landing lights. (USAF Photo)

This flight test B-2 shows its futuristic shape against a beautiful sky over Edwards Air Force Base, California. (USAF Photo)

There was also testing with engine starts being attempted in the cold weather. The tests showed no problems. The defog and heating systems in the cockpit also performed well after enduring the same icy environment.

It goes without saying that propulsion testing would play heavy in the B-2 design and development program. The ground testing was mainly accomplished by the Northrop main contractor at the Palmdale assembly location. During early testing in the mid-to-late 1980s, the testing was accomplished with the engines on stands. But during 1989, the engines were tested for the first time in their assigned locations on the first B-2 airframe. It was a huge accomplishment that would lead to the flight portion of the test program.

Although many of the details are not known, there was also considerable testing done on the B-2 structure, escape system, fuel system, landing gear, auxiliary power system, and other major components.

The crew of the first B-2 flight test, (left) Air Force Colonel Richard Crouch, Director, B-2 Combined Test Force, and Bruce Hinds, Northrop Corporation Chief Test Pilot. (USAF Photo)

This obvious flight test bird, as indicated by the black leading edge stripes, is billowing up a mist from the engine intake locations. The phenomena is caused by the weather due point condition. (USAF Photo)

The sun paints interesting patterns on this flight test B-2 as it prepares for approach into Edwards Air Force Base during the flight test program. (USAF Photo)

The ground testing also proved that the B-2 would be able to meet its designed structural endurance and planned airframe life with extensive fatigue and durability testing.

There was an air of urgency facing the program during the ground test program since with all the criticism, Air Force and prime contractor personnel felt that getting the plane into the air as soon as possible was imperative. Many also realized at the time that the chances of building the projected 132 bombers made thoughts of getting the plane flying the number one priority at the time.

Flight Testing

Even before the first B-2s flew, there was flight testing in support of the program. It consisted of a program involving a C-135, modified as an airborne testbed, which supported the B-2 program starting in 1986. The pre-B-2 flight testing involved confirmation of the bomber's Hughes radar system.

With the great confidence of the computer-designed B-2 design, there were thoughts in the late 1980s that it would not be necessary to construct a prototype for flight testing. In other words, jump directly from the computer screen to the production line.

Well, as is well known, that certainly did not happen. As it ended up, the first six B-2s would all be used for flight testing. Granted, they would all later be converted to operational configurations, but first, they played that so-important prototype role.

The first B-2 was rolled into public view on November 22, 1988, and the Air Force made the most of the world attention it received. But when all the dignitaries had departed the

This amazing 1992 photo shows an amazing scene as a pair of flight test B-2s line up for a drink from this KC-10 tanker. Refueling was a large part of the flight test process. (USAF Photo)

Hookup in this practice refueling between a flight test B-2 and a KC-10 tanker aircraft. Note how far back the refueling probe is on the crew compartment of the B-2. This test took place during 1993. (USAF Photo)

scene, the work to make it ready for flight testing resumed in earnest.

But before flight, it was necessary to conduct taxi tests, which began on July 10, 1989. The testing was accomplished at the Palmdale Airport. The initial testing involved brake efficiencies, nose wheel steering, and ground maneuvering. A portion of the taxi testing was performed with the wing trailing edge panels in the down position, which would be analogous to a flaps-down situation for a conventional aircraft. The test pilots noted that the acceleration provided by the General Electric powerplants was better than expected.

The taxi tests lasted for about a week, and then it was time to take to the air, which took place on July 21, 1989. The Air Force News Release described that first taste of flight as completely successful. The almost two-hour flight originated from the Palmdale facility and landed at Edwards Air Force Base, the site of the complete B-2 flight test program.

That first flight assessed the B-2's flying characteristics. All tests were conducted with the landing gear down in the approach configuration. The flight assessed the B-2's flying qualities. After take-off, the aircraft climbed to 10,000 feet, where functional checks of the basic subsystems were performed. The crew then brought the craft in for landing.

Crew members for this historical flight were Bruce Hinds, Northrop Corp. Chief test pilot, and Air Force Colonel Richard Couch, Director, B-2 Combined Test Force. This first B-2 would then be involved in low-observable testing and measured the bomber's stealth characteristics.

All remaining B-2 flight testing would take place at Edwards Air Force Base, with the second B-2 making its first flight test on October 19, 1990. Pilots for this flight were Northrop Test Pilot Leroy Schroeder and USAF Lt Colonel John Small. All B-2 testing at Edwards was accomplished by the 412th Test Group.

On June 4, 1992, a B-2 established another first in the flight test program when the first night flight was made. The plane's crew evaluated night flying capabilities and characteristics. In the four and one-half hour flight, the B-2 performed twilight and night takeoffs, approach patterns, and landings. The test aircraft made 11 such takeoff and landing circuits without incident.

The crew also took the plane to an altitude of 10,000 feet, where it evaluated interior cockpit lighting as darkness fell. On completing the test flight, the crew of Northrop test pilot Eric Hanson and USAF Colonel Frank Birk said there were no problems.

With long range being one of the heavy requirements on the B-2, the first series of flight testing evaluated the aircraft's capability for inflight refueling. The B-2 was tested with both KC-10 and KC-135 tanker aircraft.

During 1992, B-2s also accomplished the touchy job of night inflight refueling. The testing was accomplished at 20,000 feet, and during the first night of testing on July 2, 1992, 34 contacts with the tanker were made. The flight test aircrews were pleased with the refueling lighting and visibility of both aircraft.

Aircraft flutter tests were successfully completed, and the B-2 was deemed flutter-free for its entire operational envelope. Flutter is one of the most dangerous reactions an aircraft can experience in flight. This certification, early in the program, was a significant flight test milestone.

By 1993, all three of the six flight test B-2s were involved in the wide-ranging flight test program. It should be noted that all those flight test B-2s carried a black leading edge border easily identifying them from the production versions to follow.

Early that year, the six planes achieved the thousand hour mark with AV-2 making the mark on February 10th. In 1995, the B-2 test fleet had demonstrated the capability to operate over its entire operational altitude and speed envelope. Then, in August of the same year, the Air Force authorized the B-2 Combined Test Force to commence low-altitude Terrain Following/Terrain Avoidance test operations at altitudes as low as 500 feet. By 1996, that altitude figure would be reduced to an altitude of only 200 feet.

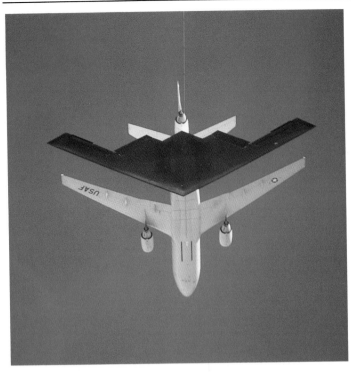

This interesting perspective of a KC-135/B-2 refueling hookup gives the impression that the B-2 is actually part of the KC-10 as the two planes are perfectly aligned. (USAF Photo)

One of the touchiest maneuvers of the refueling process is the moving in for the physical connection of the two planes. The differences in configurations makes things very interesting. (USAF Photo)

Making sure that the B-2 can fly in an icy environment is the purpose of this ice testing. This is the boom operator's view of the testing, which is emitting a liquid spray. (USAF Photo)

Weapons delivery, of course, had to be proved during the B-2 flight test program. This photo shows the release of the GAM-113 weapon during the testing. (USAF Photo)

A measure of the continued success of the flight test program was also confirmed during the period with the Secretary of Defense certifying that the original B-2 radar cross-section operational performance objectives had been successfully demonstrated from the flight test program.

Of course, the main function of a bomber aircraft is to place its weapons on target, and as such, a large portion of the flight test program addressed that capability. But there had to be separation testing on the aircraft, which means what effect the opening of the bomb bay doors and actually releasing the weapon had on the plane's stability and flight characteristics.

Early in the program, B-2s released B61 and B83 nuclear, and MK-84 conventional weapons, with no problems. By 1996, the plane had been qualified to carry and deliver up to 80 Mk 82 500-pound bombs. The newer deep-penetration weapons, the GAM-113 and B61-11, received considerable attention later in the flight test program.

This AV-4 B-2 demonstrates the drop of a B-61 bomb from the plane's Rotary Launcher during flight test, this particular operation taking place during September 1992. (USAF Photo)

Chapter 7:
Flying the B-2

It goes without saying that viewing the B-2 from a distance gives the impression that this is indeed a different animal. Lacking a characteristic fuselage and normal tail, one has to wonder about the flying characteristics of the plane.

Discussions with those who have flown the magnificent wing indicate that the B-2 has both similarities and differences in flight characteristics to wing-fuselage-tail configurations.

The first aspect of any flight is, of course, the method of entering the plane, which is accomplished on the left side of the cockpit through the crew hatch.

Former B-2 Command Pilot Lt Col (USAF Retired) Tony Grady explained the flight characteristics of the strange new bird, and he described them with one positive aspect after another.

"Sitting in the B-2 cockpit," Grady explained, "the visibility is somewhat limited. You can only see far in front of you, but you can't see the nose or the landscape close to the aircraft. The seat fits you like a glove, with the instruments further away from you than, say, in the B-52. And even though there are vertical bars in the canopy glass, you have a sense of looking through a single pane of glass.

Silky smooth and flies like a dream. That would be a composite description of those that have flown the magnificent Flying Wing machine. (Northrop Photo)

Although traversing the B-2 on the ground takes a little getting-used-to, the B-2 moves easily on the tarmac and responds to only the slightest nudge on the throttles. (USAF Photo)

The B-2 easily takes to the air as the front gear lifts from the runway. The angle-of-attack during lift-off is minimal as the rear gears lift off. (USAF Photo)

"Surprisingly, there is no inherent realization of the radical design difference in this configuration while sitting in the cockpit. Maneuvering on the ground, though, is not a problem, and it's easy to follow the taxi lines on the tarmac. One thing I found about flying this plane, though, was judging the distance from the aircraft's wingtip. There was some difficulty on the ground, but it was worse in flight.

"On the ground, this plane handles like a dream. To get it started moving on the ground, it just takes a touch on the throttles and it's underway, even with a full load of fuel. All engines are running all the time, and once the plane begins its roll, the throttles are retarded to the idle position.

"There are two basic ways we accomplished take-off, one procedure we call a rolling take-off, while the other was static, where thrust was achieved while holding the brakes.

To say that the B-2 is majestic in flight is definitely putting it mildly. As it flies by, it is certainly the king of the skies. (Phil Kunz Photo)

"On take-off, the acceleration was extremely smooth, with one of the pilots calling out the speeds while the other cross-checked his instruments. Rotation is very smooth as the plane leaves the ground at a slight angle of attack. The front gear leaves the ground first, but is quickly followed by the main gear. Angle of attack is about ten degrees-to-twelve degrees, depending on the weight of the aircraft.

"The biggest difference in the feel of the aircraft comes from its great flight control system, which operates on computer commands. The best way I can describe the feeling of the plane is to say that, 'It just felt very good in that it was responsive to precise stick inputs.'

"The complexity of the plane is unbelievable. I think one of the best examples of this is the flight control system. The rear control surfaces function as drag rudders, speed brakes, and elevons, depending on the particular flight condition." Those control surfaces, by the way, comprise about 15 percent of the total wing area.

"Being a former B-52 pilot, I quickly noticed the rigidity of the plane. The B-52's wings had a considerable flex at the tips, but with the B-2, there is very little wing movement. As such, you can feel about every bump.

"The maneuverability of the B-2 is, quite simply, awesome, with the handling during these maneuvers being more like a heavy fighter than a heavy bomber. The plane can roll very aggressively. Also, as a former FB-111 pilot, I felt similar ride quality and smoothness that I experienced in the FB-111."

With its huge wing area of more than five thousand square feet, more than twice that of the B-1B, the B-2 has considerably less wing loading. As such, the plane is less sensitive to speed and weight changes than conventional wing-and-tail heavy aircraft.

B-2 pilots will tell you that refueling the Flying Wing bomber requires extreme concentration. Here, an early B-2 in 1989 is refueled by a KC-10. (USAF Photo)

Grady indicated that both test and operational pilots were impressed with the B-2's capabilities during the often-tricky aerial refueling operation.

"When you are maneuvering the B-2 close in to the tanker, there is a smaller lag in response time in closing in on the boom than with the B-52. The biggest difference, though, with this aircraft is the bow wave that is created in front of the B-2 by its shape. This effects the rear of the tanker, especially when the B-2 has on-loaded a large volume of fuel, making the B-2 heavy, and the tanker being considerably lightened. In this situation, the rear of the tanker tends to oscillate, making it more of a challenge to keep the bomber in the air refueling position.

"Landing this plane again presents no major problems, however, landing in a strong cross-wind requires increased attention to detail to keep the wings level when close to the ground because the wingtips are so close to the ground. With the great aerodynamic efficiency of the plane, you can observe that the drag on the plane greatly increases when the landing gear is lowered. There is a significant drop in speed if the thrust is not increased in order to maintain aircraft attitude. With a Flying Wing configuration, the drag curve is very flat. If you bring this bird in too fast, it has a tendency to want to keep flying.

"All in all, this is a great airplane, and a dream to fly!" Grady explained.

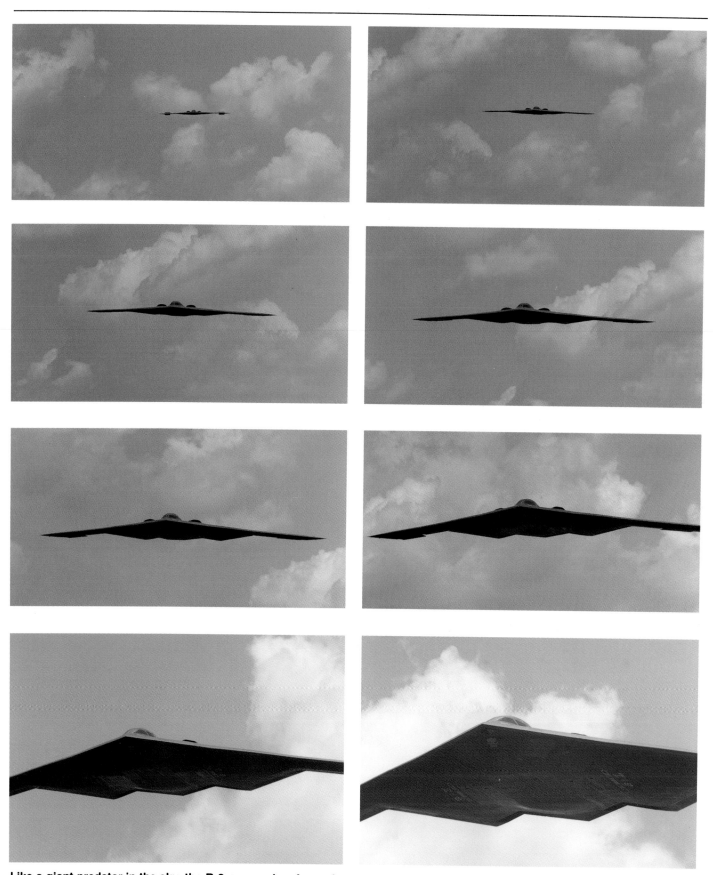

Like a giant predator in the sky, the B-2 approaches from afar and changes complexions as it nears. From its first sighting as nothing more than a tiny line in the distance until it fills the air with its huge size and sawtoothed rear end. (Sequence by Phil Kunz)

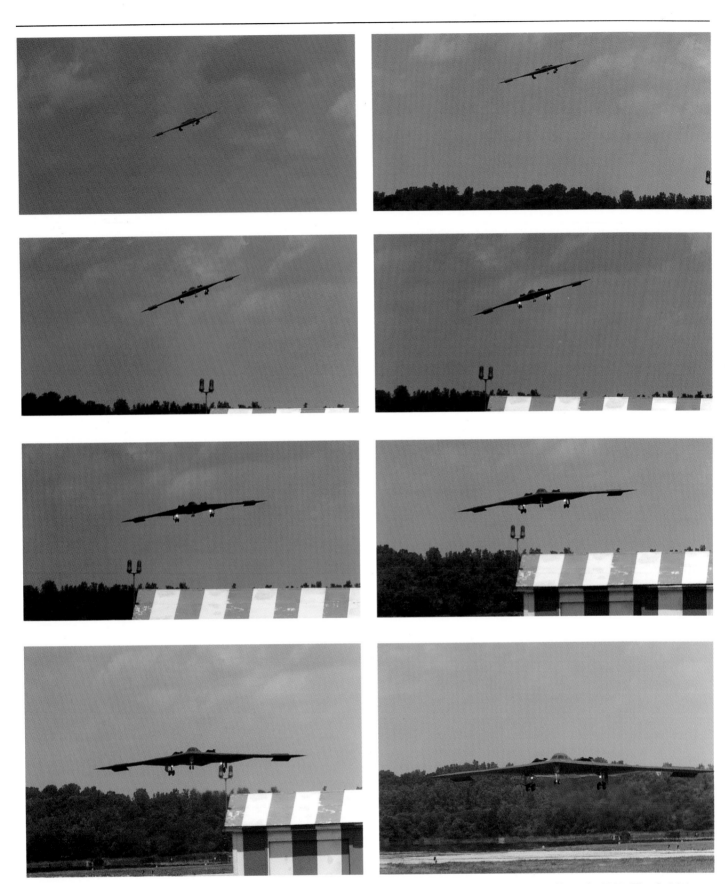

This sequence shows a graceful landing sequence of a B-2 bomber at Wright Patterson Air Force Base in 1997. The Spirit has a slight nose-up attitude when the rear gears touch the ground raising the expected twin puffs of smoke as the tires are brought up to speed. (Phil Kunz Photo Sequence)

Chapter 8:
B-2 Operational Service

Of course, it does not make much sense to design and develop a potent new weapon system without the logistical and operational infrastructure to support it. With the B-2 bomber, that planning began early, and when the plane was ready to be deployed, the support structure was in place.

Base Construction Implications
Whiteman Air Force Base, located some 45 miles southeast of Kansas City, Missouri, was selected as the site. Interestingly, it had been over three decades since the base had housed strategic bomber aircraft with B-47s in place up until the mid-1960s.

With its illustrious past, it seemed right that the 509th Bomb Wing be given the revolutionary bomber. It was the 509th that was created in December 1944 with the mission of dropping the first atomic bomb on Japan.

Initial appropriations for construction at Whiteman occurred in Fiscal Year 1988, with the primary focus being the building of new hangars to house the new bombers, along with a combat crew training facility. Initial plans called for a

The lack of a black leading edge on the wing indicates that this is an operational model, actually one of the first, being the AV-8 flight vehicle. (USAF Photo)

The fact of American deterrence in the next century could well be visualized by the shape of the mighty B-2 bomber shown here with the sun cascading off its futuristic lines. (USAF Photo)

This operational B-2 casts a painter's look with the sun highlighting its graceful crew compartment and stealthy engine intakes. (USAF Photo)

Crew members pass each other on the taxi-way at Whiteman Air Force Base during 1994 time period. (Northrop Photo)

covered maintenance area for each bomber. It would all be the home for the lone B-2 organization, the 509th Bombardment Wing. Other new B-2 associated facilities, which were constructed at the Whiteman complex, included a new maintenance control complex, a mission operations center, a radar approach control facility, and aircraft support equipment and parts storage facilities.

Maintenance is a heavy player with the B-2, with the Oklahoma City Air Logistics Center at Tinker Air Force Base, Oklahoma, being denoted as the prime B-2 maintenance facility. The first B-2 arrived at Whiteman in December 1993, with others following through the years. With such a new weapon system, it was starting with a clean slate as far as the operation would evolve. General Ronald Marcotte, the first commander, explained that it was a situation of crawl, walk, and then run in bringing the operation to fruition.

The situations presented by the new plane would be different from bombers of the past, due to its extreme complexity. In fact, when operations first began, the B-2 was still in its flight test program. But with all those considerations, the unit moved quickly and attended its first Red Flag exercise a year ahead of schedule.

Needless to say, the personnel that were selected to be a part of this first experience with the B-2 were the best available. Each candidate had to have a spotless record and a high recommendation, other endorsements, and thousands of hours of flight time. It worked out that many of the pilots selected had considerable combat experience. Another interesting criteria used in the selection process was maturity, resulting in an older group of crew members.

Coming aboard with the 509th required a six-month training period, which included flying, but also allowed the crew

This B-2 rests in a Whiteman maintenance hanger under the careful surveillance of a technician. Note the clam-shell doors of the weapons bay in the open position. (Northrop Photo)

members to participate in the formation of tactics for the plane. Many of those first pilots were involved in Desert Storm, and that experience proved invaluable in this undertaking. These first crews were also valuable contributors to the first Dash-1 procedures manual.

Flight training involves flying to bomb ranges in Utah, Wisconsin, and Kansas. About one of five of the weapons dropped was a live weapon.

The extensive facilities at Whiteman were built with longevity in mind, as it has been stated that the B-2 could last as long as four or five decades. There is also additional space at the base for additional B-2 facilities should they be required.

1997 was a momentous year for the 509th, with the Wing achieving Limited Operational Capability (LOC). The LOC designation allowed the unit to be available to theater commanders in a conventional role. Official Initial Operational Capability (IOC) occurred shortly thereafter. Throughout the year, the planes of the fleet continued to be updated. In fact, at the end of 1997, the 509th received the third Block 30 B-2, actually AV-2, which had originally been one of the initial flight test vehicles. The first Block 30 plane, the "Spirit of Pennsylvania," arrived in August, while the second, "The Spirit of Louisiana," arrived in November. It was planned that there will be more Block 30 versions arriving in 1998.

A part of the 509th are a pair of Bomb Squadrons, the 393rd and the 715th. Planning called for the 393rd to reach Full Operational Capability (FOC) in the 1997 time period, with the 715th planned for the 1999 time period. Planning called for each squadron to field eight B-2s, which accounted for 16 of the 21 production aircraft.

The AV-8 was the first production B-2 to arrive at Whiteman Air Force Base. (USAF Photo)

Operations

The unit was ready for its new status, with well over one thousand sorties having been accomplished through December of 1996. Included in that impressive title were participation in Red Flag exercises and a number of overseas missions. In 1997, the 509th B-2s participated in Global Power operations.

A record-setting flight took place in May 1997 when a 509th B-2 took off from Whiteman and flew a 30-hour round-trip mission over RAF Mildenhall, UK, and then included simulated bomb runs over mid-America. Another milestone occurred the same month when a B-2 dropped a Global Positioning System-Aided Munition over Edwards Air Force Base. The mission was the first double dropping of the conventional GAM-113 with the B-2.

The capability was proved when the B-2s delivered 16 GAMs against the same number of targets with hits on every target. All the weapons were released from above 40,000 feet, with distances of up to seven miles from the targets.

In November 1997, the 509th again illustrated the awesome B-2 capabilities in support of an air power demonstration at Eglin Air Force Base, Florida. The B-2 successfully tested the plane's automated mission planning system for dropping two weapon systems, including the JDAM weapon. This was the first time the munition was employed using operational maintainers, loaders, crews, and an operational bomber. The mission was also the first time a B-2 crew performed in-flight re-planning of a target.

It was also announced in late 1997 that the B-2 could now fly 600 hours between planned maintenance inspections, which according to the Air Force is the longest for any of its aircraft systems. The new maintenance intervals were increased from the previous 200 hour intervals. During each

AV-11 awaits in its maintenance hanger while one of its brothers takes to the air in the background at Whiteman Air Force Base. (USAF Photo)

inspection, more than 70 panels are removed, along with the inspection of some 1,200 parts. During this inspection, the B-2 is also checked for corrosion and structural integrity.

In early 1997, the 509th experienced a minor set-back when a standdown was ordered of all training missions. The reason was a fracture on an engine-to-accessory drive system shaft in one of the bombers. The stand-down was canceled when a satisfactory solution to the problem was accomplished by flight testing.

Attack Considerations

The stealth capabilities of the B-2 make one of its missions very obvious, that being to be a lead element in the attack of an enemy target on the first night of any future confrontation. Its mission would also be expected to attack the most important and probably the most heavily-defended targets.

In a 1992 statement, the B-2's warfighting capabilities were described as "being able to make the initial penetration into the target area, neutralizing the defenses and allowing the less-stealthy systems to then operate."

A USAF report at the time described the B-2 as a pivotal tool for bringing about the rapid destruction of an enemy. The capability was to be accomplished "from U.S. bases attacking the enemy nerve center of enemy capabilities alone and at will."

The B-2 will operate in a so-called "Direct Attack" mode, which it is argued is a much more economical method of attack than the current stand-off mode as performed by the venerable B-1 and B-52H bombers.

An interesting cost comparison shows that at one million dollars per copy, cruise missiles are at least 50 times more expensive than the B-2's direct-attack munitions. It also takes

This 1997 photo provides a rear view when three operational B-2s are in sight. Two are on the move on the taxiways while a third can be sighted in its maintenance hanger in the shadows. (USAF Photo)

two-to-four non-stealthy B-52s carrying a total of 32 cruise missiles to match the combat power of a single B-2 sortie.

Used instead of cruise missiles, the B-2 saves over $31 million on weapons with each strike. Also, the B-2 weapons normally carry two thousand pound warheads as compared to half that weight on the cruise missiles. Of course, at this time, it is all theoretical, and will have to be verified someday in battle. Hopefully, it will never have to be proven.

An operational B-2 rolls to takeoff position behind a maintenance hanger where another B-2 resides. (USAF Photo)

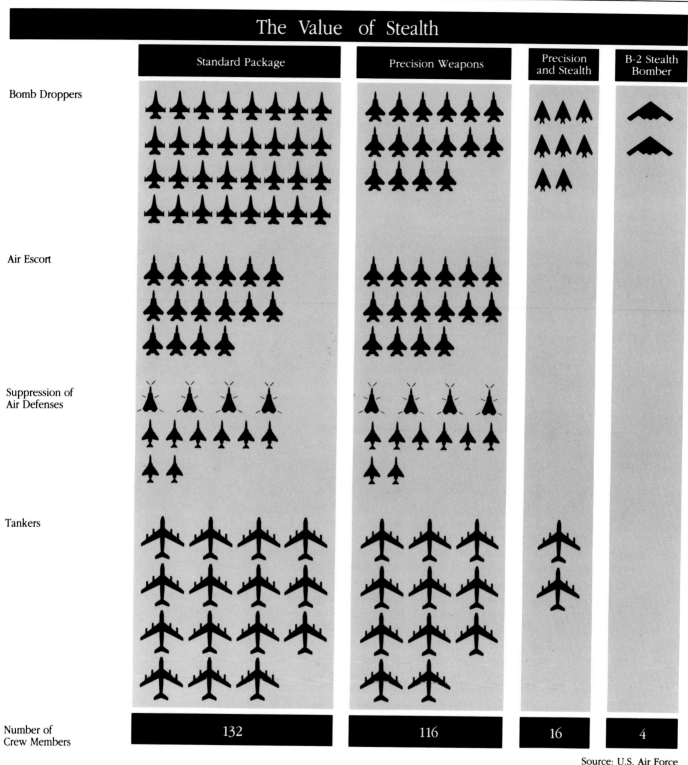

The warfighting capabilities of the B-2 are shown by this chart. Even though the B-2s are few in number, their individual capabilities make them seem like many more. (USAF Drawing)

An amazing photo at Whiteman Air Force Base where three operational B-2s are on the move to take-off positions. Pilots explain that the B-2 is an extremely responsive aircraft to maneuver on the ground. (USAF Photo)

A rare underneath view of the B-2 clearly shows the landing lights turned on, each pair of lights being located with each of the three landing gear units. (USAF Photo)

Also from the publisher

LOCKHEED F-117 NIGHTHAWK: AN ILLUSTRATED HISTORY OF THE STEALTH FIGHTER

Bill Holder & Mike Wallace

The F-117 was probably the most secret aircraft ever developed. This new book covers the technical and operational aspects of the Nighthawk from its initial use over Panama through its shining moment during Operation Desert Storm.

Size: 8 1/2" x 11"
over 120 color & b/w photographs
64 pages, soft cover
ISBN: 0-7643-0067-9

$19.95

McDONNELL-DOUGLAS F-15 EAGLE: A PHOTO CHRONICLE
Bill Holder & Mike Wallace

Photo chronicle covers the F-15 Eagle from planning and development to its success in Operation Desert Storm and post-Desert Storm. All types are covered, including foreign – Israel, Japan and Saudi Arabia.
Size: 8 1/2" x 11"
over 150 color & b/w photographs
88 pages,   soft cover
ISBN: 0-88740-662-9 $19.95